First-Year Physics for Radiographers

This book is dedicated to the memory of Juliette Burla, Superintendent Radiographer and Tutor in the Department of Radiotherapy, the General Infirmary at Leeds, whose sudden death on 13th February 1969 deprived student radiographers of a wise counsellor.

First-Year Physics for Radiographers

SECOND EDITION

George A. Hay
M.Sc.
Senior Lecturer in Medical Physics in the University of Leeds, the General Infirmary at Leeds

and

Donald Hughes
B.Sc., Ph.D., C.Eng., F.Inst.P., M.I.E.E.
Radiation Protection Officer in the University of Leeds

BAILLIÈRE TINDALL · LONDON

A BAILLIÈRE TINDALL book
published by Cassell Ltd
35 Red Lion Square, London WC1R 4SG
and at Sydney, Auckland, Toronto, Johannesburg
an affiliate of
Macmillan Publishing Co. Inc.
New York

First published 1972
   Reprinted 1975, 1976
Second edition 1978

ISBN 0 7020 0692 0

Printed by Cox & Wyman Ltd, London, Fakenham and Reading

British Library Cataloguing in Publication Data

Hay, George A
   First year physics for radiographers. — 2nd ed
   1. Physics 2. Radiography
   I. Title II. Hughes, Donald, 1931
   530'.02'4616   QC28

ISBN 0–7020–0692–0

Preface to the Second Edition

The Second Edition covers the Physics Syllabus (1976 Edition) for Part I of the Examination for the Diploma of the College of Radiographers. It has been extended or modified in only a few places, notably in the section on solid-state rectifiers and in the emphasis on expressing X-ray spectra in terms of photon energy rather than of wavelength. A section has been included also on shapes and fine details in the X-ray image, referring particularly to unsharpness in radiological systems. The new S.I. units of absorbed dose and of activity, the gray and the becquerel, have been mentioned.

We hope that the book will continue to prove of value to students.

March 1978
G.A.H.
D.H.

Preface to the First Edition

This book covers the Physics Syllabus (1971 Edition) for Part I of the Examination for the Diploma of the Society of Radiographers. It was written because no single textbook could be found that just covered the course; most books that reached the required standard on some topics, for example on alternating currents, contained an unnecessary amount of academic detail on many others. Some attempt has been made to anticipate possible future changes, notably in the chapter on radioactivity and in the adoption of S.I. units (Système International d'Unités).

The text is intended to be almost completely self-explanatory; some of it has already been used with success in teaching by the following method. The student first reads a topic selected by the lecturer; this is then discussed in detail in class. Finally the student uses the text as revision notes, with the assurance that no irrelevant material is included. Indeed, the text is based on revision notes given to student radiographers over a period of many years; the main development has been a very comprehensive system of cross-referencing that is designed to assist comprehension. (The student is advised to read the 'instructions' on page vii.)

We are indebted to Mr G. W. Reed, Senior Lecturer in Medical Physics in the University of Leeds, for his invaluable criticism and advice. We also wish to thank Mr H. B. Bentley, Principal of the School of Radiography, The General Infirmary at Leeds, and Mr E. Naylor, Principal of the School of Radiography, Bradford Royal Infirmary, who read and criticized some of the early drafts. The book would not have appeared without the painstaking typing of Miss Claire Morgan.

February 1972   G.A.H.
                                                                                         D.H.

To the Student

This book contains a continuous and logical account of the physics you must study. If you read it from beginning to end, and if you understand and remember all you have read, you may feel confident that you can pass the examination – if you can write it down adequately!

Do not allow anything to go by without understanding it; make use of the many cross-references (referring *backwards only* on the first reading but both *forwards and backwards* when revising later) and ask advice if anything is not clear. Memorize the definitions and understand what they mean. Your teacher will help by discussing the various topics; he will also provide you with practice in working simple numerical calculations and in writing answers to examination questions. There is little in the book you needn't know; there is little outside the book you must know!

Good luck!

Contents

CONTENTS

xi

1 General physics

1.1 INTRODUCTION

1.1.1 **Physics** is a science which is concerned with the study of two concepts, **matter** and **energy**, and how they interact with each other.

1.1.2 **Matter**, which may be **solid**, **liquid** or **gas**, is the physicist's name for the material of which everything in the universe is composed. Common examples are copper, rubber, water and air. Matter is composed of sub-microscopic units called **atoms** or **molecules**. Atoms often combine with each other to form molecules; the distinction between atoms and molecules will be made clear in Chapter 2.

In general, the physicist studies the behaviour of matter only when it does not change into other chemical forms of matter. For example, the laws governing the movements of bodies through space and the manner in which atoms are constructed come within the province of physics. On the other hand, the way in which atoms of hydrogen and oxygen combine to form molecules of water, or sulphur burns to form sulphur dioxide, comes within the province of **chemistry**.

1.1.3 **Energy** is a concept which is difficult to define concisely but which is relatively easy to understand. A common definition is

DEFINITION   **Energy is ability to do work.**

This definition resembles closely our everyday use of the word. For example, one may wake one fine morning and think 'I feel full of energy' which generally means that one is able to do a lot of hard *work* (or *play,* which has the same significance for this purpose!).

Definitions in physics are seldom so consistent with everyday meanings as this one. Most of us, for example, would consider that we would be doing quite a lot of work while standing still holding a heavy suitcase. Here, however, we should be doing no 'work' according to its definition in physics (1.3.4). It often happens that a word in physics means something quite different from its everyday meaning; any resulting confusion must be guarded against, and it is therefore very important to define words precisely.

The full meaning of the concept of energy will emerge more clearly in later sections. It is sufficient to say now that there are several different forms of energy: mechanical, electrical, chemical, heat, light, X-ray, etc. All forms of energy can be converted more or less easily into other forms; for example, *electrical* energy is easily converted into *heat* energy in the domestic electric fire, and *heat* is less easily converted into *mechanical* energy in the steam engine.

1.1.4 The conservation laws. Most forms of matter can be converted into other forms, and all forms of energy into all other forms. Up to the beginning of this century, these two types of conversion process were described by two important laws which were called respectively **the law of conservation of matter** and **the law of conservation of energy.** These laws stated quite simply that

DEFINITION   **Matter can be neither created nor destroyed,** and

DEFINITION   **Energy can be neither created nor destroyed.**

According to these laws, if we converted a certain amount of matter (or energy) into another form, we should always finish with exactly the same amount of matter (or energy) in the new form.

Originally these two laws were thought to be absolutely true and independent of each other. Then physicists discovered that changes in the structure of the atoms of matter were possible, resulting in the release of large amounts of energy. These so-called **nuclear** changes occur spontaneously in radioactivity (Chapter 16); in the late 1930s it was discovered how to produce them at will, culminating in the development of nuclear weapons and later of nuclear power stations. In these nuclear changes, matter is actually being destroyed and is being converted into energy. However, this new type of process need not make the conservation laws untrue if the two laws are considered together as one. For example, one may say that matter may be converted into an equivalent amount *either* of another kind of matter (e.g. solid to liquid) *or* of some form of energy. Then the total amount of matter plus energy in a system remains constant.

This new concept may be formally expressed in a combined conservation law:

DEFINITION   **The total amount of matter and energy in an isolated system is constant.**

The word 'isolated' means that nothing is being added to or taken away from the system; there is no 'outside interference' with it.

1.2 MEASUREMENTS AND UNITS

1.2.1 Qualitative and quantitative descriptions. In studying matter and energy and their various properties, physicists are not content merely with describing *what* happens in different circumstances, i.e. with a **qualitative** description. They insist on the importance of stating also *how much* of the particular effect occurs, i.e. in terms of a **quantitative** description. This principle can be illustrated from radiology itself, which is a complex application of physical principles: it is very important to be able to measure the X-ray exposure to a patient or to a radiograph so that the patient will not be harmed or so that the radiograph will be satisfactory. We shall therefore now discuss the important subject of measurement and the units in which the answer can be expressed.

1.2.2 Dimensions to be measured. Despite the wide variety of quantities in physics which must be measured, for example volume, velocity, density, force and voltage, it is remarkable, and fortunate, that for most purposes only three basic types of measurement are necessary. These **dimensions**, as they are called, are **mass, length** and **time** (M, L and T). For example, mass is a direct measure of quantity of matter (1.3.3), length cubed (L^3) expresses **volume** and length divided by time (LT^{-1}) is **velocity**.* An important quantity, to be used on a number of occasions in this book, is **density**. This is a measure of

*If the word *velocity* is unfamiliar, it can be regarded for the time being as synonymous with the more familiar word *speed*. The precise difference between the two will be explained in section 1.3.2.

the quantity of matter per unit volume, and is therefore calculated by dividing the mass of a given body by its volume (ML^{-3}). For example, the density of water is about 10^3 kg m^{-3}, of aluminium about $2 \cdot 7 \times 10^3$ kg m^{-3} and of lead about $1 \cdot 1 \times 10^4$ kg m^{-3}. Because of the way in which all these quantities may be **derived**, nearly all measurements in physics (and in related subjects such as engineering and astronomy) are made in terms of the three fundamental dimensions.

1.2.3 Units. It is impossible to measure quantities without having agreed units at our disposal; for example, how could one express the distance from London to Paris without having first defined the mile or the kilometre? Similarly, in physics, the quantities mass, length and time are measured in standard units. In the greater part of the book we shall adopt the recently standardized S.I. (Système International) units; the basic units are the **metre** (m), the **kilogram** (kg) and the **second** (s). Certain other units are derived from these. However, in special cases where older units are particularly widely used and understood, for example the **calorie** for heat, we shall depart from the S.I. units. The way in which units in general are used will become apparent in the remainder of the book.

1.3 FORCE, WORK AND ENERGY

1.3.1 Force is a concept which is familiar in an everyday sense to most of us. Its meaning in physics, however, is rather more restricted; it is expressed in the definition:

DEFINITION　**Force produces or tends to produce movement in a body.**

For example, if we push against a small box on the floor, movement is produced. If we push against a brick wall, we do *not* produce movement, but we are still applying a **force** to the wall. We then only *tend* to produce movement. If we push with a bulldozer, of course, we *shall* produce movement!

In the example of the box, to produce continuous movement we must apply a continuous force, i.e. we must keep on pushing. This is because of friction between the box and the floor. If we could take the box into outer space, where there are no disturbing factors, and give it a single, short push (i.e. we apply a temporary force to it) it will acquire a certain *velocity*. It will go on moving in a straight line at that velocity for ever unless other forces happen to act on it (such as, for example, the attracting influence of a planet). A given force, acting for a given period on a body which is free from all other forces, imparts to it a constant velocity. If the body contains more matter, i.e. it has a larger mass, the velocity so produced will be less. If the force acts continuously, instead of being limited to a certain period, the velocity of the body will go on increasing, i.e. it will accelerate.

These ideas enable us to define a standard or unit of force in terms of the basic units of mass, length and time. Suppose we have a body of unit mass of 1 kilogram (kg), and we apply a force to it such that at the end of the first second it has accelerated to a velocity of 1 metre per second (1 m s^{-1}), at the end of the *next* second to 2 metres per second, etc., we say that the body has an acceleration of 1 metre per second per second (1 m s^{-2}), that is an increase of velocity of 1 m s^{-1} in every second. The force that would produce this result is called one newton (N).

DEFINITION   The unit of force, called the newton, is that force which, when applied to a body having a mass of one kilogram, gives it an acceleration of one metre per second per second.

The newton is called a derived unit because it is *derived* by *definition* from the basic units.

1.3.2   Scalar and vector quantities. The concepts of mass and force are examples of two important kinds of quantity which must be distinguished clearly. Mass has no idea of direction associated with it, and is called a scalar quantity, whereas force may be upwards, sideways, downwards, etc., and *has* direction; it is called a vector quantity.

DEFINITION   A scalar quantity has magnitude (size) only.

DEFINITION   A vector quantity has direction as well as magnitude.

We shall meet many other examples of scalars and vectors throughout the book, but important examples that we have already met are speed and velocity. In section 1.2.2 (footnote: p. 3) it was stated that velocity could be temporarily regarded as synonymous with speed. However, *speed* is a *scalar* quantity and *velocity* a *vector* quantity. For example, we may say that we are driving at a *speed* of 50 miles per hour, but this does not tell us *in what direction* we are driving. We may further say that we are driving at a velocity of 50m.p.h. *due north*, thus including the information about *direction* which is an essential part of a vector quantity.

If then we drive round a bend so that we are going *due east* at 50m.p.h., we have changed *direction* and therefore *velocity*, but our *speed* has remained constant. The change of velocity in going round the bend is an **acceleration** just as much as would be a change of speed from 50 to 60m.p.h.

1 3.3   Mass and weight; gravitation. In section 1.2.2, mass was stated to be a measure of quantity of matter. However, the question arises, 'How can we measure mass in practice?' It could be done by applying a known force and measuring the resulting acceleration (1.3.1), but this would be highly inconvenient. Instead, the concept of **weight** is used. In practice, for example on holiday on the Continent, if we wish to buy a 'mass' of 1 kg of meat, the amount is determined by *weighing*.

The concept of weight depends on the fact that all bodies, whatever their mass, exert on each other a force of attraction, called **gravitation**. The force between two comparatively small masses, such as two bodies each of 1 kg, is very small and difficult to measure. However, the earth has such a large mass that it exerts a considerable force on any other body near it. This force, known as the **force of gravity**, is directed towards the centre of the earth, and thus always appears to be acting 'downwards'. It is the force which tends to make all bodies fall to the earth from a height, and which Galileo was investigating when he did his well known experiment from the Leaning Tower of Pisa.

The force of gravity exerted on a body can easily be measured, for example by measuring the stretch of a spring in the familiar spring balance. The force *could* be measured in newtons. However, for the sake of simplicity of expression the force which gravity exerts on a *mass* of one kilogram is called one **kilogram-weight** (often abbreviated to **kilogram**). Hence, in practice, mass is measured by measuring the gravitational force in kilograms-weight exerted on the body; the mass in kilograms is then numerically equal to the body's weight. For example, a mass of 3 kg weighs 3 kg wt; a mass of 100 kg weighs 100 kg wt. For reference, 1 kg wt = 9·81 newtons.

1.3.4   Work has been mentioned previously (1.1.3) but its meaning in physics is somewhat more restricted than in everyday life. Figure 1.1 shows a cliff overhanging the sea (which is assumed to be dead calm and free from

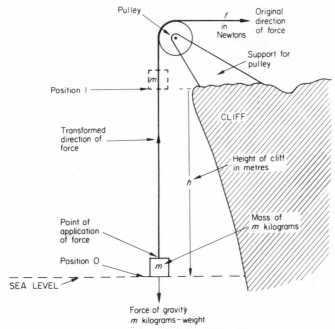

Fig. 1.1   An experiment to illustrate work, potential energy and kinetic energy.

tides!); the top of the cliff is *h* metres above sea-level. A body of mass *m* kg is suspended by a rope and pulley system; suppose initially the body is just in contact with the sea (position 0).

If we wished to raise the body, say to position 1, level with the cliff-top, we should have to do **work** by pulling *horizontally* on the rope at the cliff-top. The pulley merely makes it easier to pull by altering the direction of the rope; the true direction of the force on the body is *vertically upward.* Let us call the force we must exert *f* newtons. Commonsense tells us that we should have to walk a distance *h* metres, pulling the rope, to lift the body from position 0 to position 1.

Then the work (*w*) that we should have to do to raise the body is equal to the force times the distance moved by the point of application of the force, in this case

$$w = fh.$$ Eq. 1.1

The unit of work (and **energy**, see 1.3.5) is the **joule (J).**

DEFINITION   The unit of work, called the joule, is the work done when the point of application of a force of one newton moves through a distance of one metre in the direction of the force.

(The phrase 'in the direction of the force' is necessary because sometimes the body does not move in the direction of the force; then an additional factor must be introduced which we shall not discuss.)

For work to be done it is not necessary for the movement to be *vertical;* it can be in any direction, and need not involve gravity. The example in Fig. 1.1 (1.3.4) has been so chosen that it will illustrate other important principles in the following sections.

1.3.5   Energy: potential and kinetic. In Fig. 1.1, when we have raised the body to position 1 by expending work equal to fh joules on it, we must have changed its state in some way. Obviously it is h metres higher than before, but what does this really mean? What has happened to the work we have so laboriously put into the body? The fact is that it has been converted into **energy**. Because we are dealing with a mechanical phenomenon (rather than heat, electricity, etc.) it is called **mechanical energy**; because it is energy of *position*, it is also called **potential energy**.

How can we recover this energy from the body? We can do so simply by allowing the body to fall back to position 0. If we do this slowly, in a controlled manner, we can make use of the energy in any desired way, for example by attaching something to the rope and by allowing the force f to move again through the distance h, so doing an amount of work $w = fh$ at the cliff-top. Thus we see that potential energy satisfies the definition in section 1.1.3: 'Energy is ability to do work'. Alternatively, we could let the body fall rapidly, so that at the instant when it reached position 0 it could be made to give up its energy and to do useful work such as driving a pile into the sea-bed. In this case, as the body reaches the position 0 it still possesses all its energy; here the *potential energy* has been converted into energy of *movement*, called **kinetic energy**. As we have just shown, both potential energy and kinetic energy are mechanical in nature and are both convertible into work. Like work, energy is measured in joules (1.3.4).

Before leaving Fig. 1.1, we must discuss further the ideas of **height** or **level** which are of very great importance in physics. A body naturally falls from a greater height to a lesser height; water flows from a higher level to a lower level. These things happen because the matter that moves has a greater potential energy at the greater height or level. We shall see in later sections and chapters how similar phenomena occur in both heat and electricity, and these are much more easily understood if the mechanical ideas embodied in Fig. 1.1 are clear.

Another idea concerns the *measurement* of height or level; in Fig. 1.1, *h* represents in metres only the *difference* in height between the sea and the cliff-top. Of course, on earth it is usual to measure heights from sea-level, but this is a mere convention. We cannot say, for example, that the body in Fig. 1.1 has *zero* potential energy at position 0, because if we let it fall further (and if it did not float) it would sink to an even lower level and have an even lower potential energy. Thus heights, levels, potential energy and kinetic energy as well as a lot of other quantities are only *relative* to some agreed **zero**. It is important that the zero of the quantity be wisely defined otherwise difficulties might arise in subsequent applications of the measurements.

1.4 TEMPERATURE AND HEAT

1.4.1 Temperature. We all know the difference between a *hot* body and a *cold* body. If they are not *too* hot or cold, we can tell the difference by touch; we can even, to a certain extent, detect intermediate stages of 'hotness' and 'coldness'. What precisely is the physical meaning of this?

In section 1.1.2, matter was described as being composed of very small particles called **atoms** or **molecules**. Besides being so small that they are quite invisible, even under the most powerful microscope, these particles are in incessant movement to and fro. In solids, this movement is *regular* and is called **vibration** (rather like the vibration of a guitar string). In liquids and gases it is *haphazard,* and is normally called **random movement.** Whatever the type of movement, it results in the atoms or molecules possessing kinetic energy (1.3.5). It is this kinetic energy which has been found to be responsible for the hotness and coldness, or **temperature**, of a body. If the atoms or molecules are in vigorous movement, and therefore have a *high* energy, the body is said to have a **high temperature**. If the atoms or molecules have only·*low* energy, the body has a **low temperature.**

The effects of *changes* of temperature on matter are numerous and will be only mentioned here. With rise of temperature, for example by **heating** in a gas flame, nearly all matter **expands** (occupies a greater volume). This is because the more vigorous movement of the atoms or molecules tends to a mutual 'pushing apart'. If a *solid* is heated, at a certain temperature (known as the **melting point**) it **melts** and changes into a *liquid.* If the *liquid* is then heated, at a higher temperature (known as the **boiling point**) it **boils** and changes into a *vapour*. Both types of change are very familiar in the transitions from ice to water and from water to steam. In fact, the temperatures at which ice melts and water boils have supplied the two principal points for the **Celsius** or **Centigrade** scale of temperature. The zero ($0°$ C) of the scale is the melting point of ice, and one hundred degrees ($100°$ C) on the scale is the boiling point of water (both under certain specified conditions). The interval between the two is divided into 100 degrees ($100°$).

Temperature is commonly measured by instruments called thermometers; in these, a change of temperature causes the expansion or contraction of a liquid (usually mercury) in a tube. The volume of the liquid in the tube is indicated by its length on a calibrated scale.

If matter is cooled to lower and lower temperatures, the kinetic energy of the atoms or molecules becomes less and less. One can in fact imagine a stage at which all movement ceases. This temperature, which has never been quite attained in practice, is called the **absolute zero** of temperature. Its value is about $-273°$ Celsius.

1.4.2 Heat. Consider two blocks of metal, for example copper, one at a *high* temperature and the other at a *low* temperature. If they are placed in close contact, experience tells us that the hot body will become cooler and the cold one warmer, until finally they are both at the same temperature. We can explain this in terms of atomic or molecular energy by saying that the more vigorous particle vibrations of the hot body transfer some of their energy to the particles of the cold body. In other words, there is a **flow of energy** from one body to the other. This particular form of energy is called **heat,** and the way in which it is thus transferred through solid matter is called **conduction** (1.4.3).

Being a kind of kinetic energy, it would be natural to measure the quantity of heat in **joules** (1.3.4), and this in fact is the S.I. unit of heat. However, the existence of heat as a kind of 'substance' which appeared to 'flow' through matter has been known for a very long time. Before the properties of atoms and molecules were understood, therefore, heat was regarded as a weightless fluid, with special properties of its own, that could be made to flow into a body and to raise its temperature.

Accordingly, a special unit of heat was derived, called the **calorie** (cal). Despite the adoption of the S.I. units, this special unit is still in very general use, and it will be employed throughout this book wherever heat is referred to.

DEFINITION   **The calorie is the amount of heat which will raise the temperature of one gram of water by one degree Celsius.**

One calorie is the equivalent of about 4·2 joules.

It is most important to appreciate the difference between *heat* and *temperature*. This difference can perhaps be made clear by pointing out the similarity, or **analogy**, between flow of heat and flow of water. *Water*, whose quantity may be measured in units of volume, flows from a place at a *high* level (or height) to a place at a *low* level (or height). Heat, whose quantity is measured in calories, flows from a body at a *high* temperature to a body at a *low* temperature. This *temperature* may be regarded, so far as heat is concerned, very roughly in the same way as *level* or *height*. Such similarities or analogies very often help us to understand new concepts in terms of ideas with which we are already familiar.

From the definition of the calorie, given above, we can reason by simple proportion that, for example, 5 calories will be needed to raise the temperature of 1 gram of water by 5 degrees Celsius, or 20 calories, 4 grams of water through 5 degrees Celsius. If m (g), is the mass of water, $t(^\circ C)$ the rise in temperature, and H (cal) the amount of heat, then

$$H = mt. \qquad\qquad \text{Eq. 1.2}$$

But what about substances other than water? Do they behave similarly, or do they require different amounts of heat to raise their temperature?

It is interesting that, for a given rise of temperature, a given mass of water will hold more heat than an equal mass of nearly all other substances. This may be expressed in terms of the **specific heat** (s):

DEFINITION   **The specific heat of a substance is the amount of heat required to raise the temperature of one gram of the substance by one degree Celsius.**

According to the above, the specific heat of water is $1\cdot0$ cal/g$^\circ$C and of all other substances is less than $1\cdot0$. For example, s for copper is about $0\cdot1$ cal/g$^\circ$C.

To take account of different specific heats, Eq. 1.2 must then be expanded to:

$$H = mst. \qquad\qquad \text{Eq. 1.3}$$

For example, the heat required t $\jmath$ raise the temperature of 4 g of *copper* by 5° C is $H = 4 \times 0\cdot1 \times 5 = 2$ cal (compare with 20 cal for water).

Sometimes we are interested in the behaviour not of one gram of the substance, but of a body of any mass. Then we speak of the **thermal capacity** or **heat capacity** of the body.

DEFINITION **The thermal capacity of a body is the amount of heat required to raise its temperature by one degree Celsius.**

This definition is interesting because it has an exact and very important counterpart in electricity (3.4.1).

1.4.3 Conduction, convection and radiation are the three ways in which heat energy can be transferred from one place to another. They are very important in radiology, because the design of an X-ray tube depends for its success largely on the way in which these processes occur in the tube (11.3.4).

We have already seen (1.4.2) how heat energy is **conducted** through *solids* by the transference of the vibrational energy of atoms or molecules through the solid. Different solids behave differently in this respect; metals in general are *good* **conductors** of heat, and non-metals are *bad* conductors. Among the metals, silver and copper are the best heat conductors, and iron and tungsten are not so good.

Conduction also takes place in *liquids* and in *gases,* but it is very difficult to detect, for two reasons. One is that they are very bad conductors, but the more important reason is that another transfer process called **convection** takes place which completely obscures the small amount of conduction.

If we put a saucepan of cold water on a very low gas flame, soon we shall see currents of water rising from the heated part of the pan. The effect is clearer if a few crystals of a soluble coloured material (such as potassium permanganate) are carefully placed at the bottom of the pan. This process, called **convection**, occurs because the heated water at the bottom expands and becomes less dense; it therefore rises to the top, *carrying the heat with it.* Cold water from the top sinks to take its place. It is well known that the air near the ceiling of a heated room is much hotter than that near the floor; people will say 'heat rises', but in fact it is the *hot air* which rises, carrying the heat with it.

The third type of heat transfer process is called **radiation**. This is the way in which we receive heat energy from a hot fire or from an electric radiator.

Conduction cannot be involved, because there is no solid matter between us and the fire. Convection cannot transfer the heat, because air heated by the fire, being less dense, will travel upwards, not sideways. In fact, **radiated heat** can be shown to travel through space where no matter exists at all. The commonest example of this process is the way in which heat reaches us from the sun; it is well known that most of the distance between the earth and the sun is almost empty space.

It seems that the vibrating atoms or molecules of the hot body are in some way sending out *rays* which either carry energy or consist of energy. These rays are similar in nature to light rays, X rays, radio waves, etc., and have been given the name **electromagnetic waves** or **electromagnetic radiation**. Their nature and properties will be fully discussed in Chapter 9.

Heat **radiation**, while being quite different from heat conduction, resembles it in that heat is radiated only from a body at high temperature to a body at low temperature. For example, in summer, when the sun is shining brightly, we can feel that we are receiving heat from the sun; this is because we are at a lower temperature than the sun! In winter, however, when the sun is obscured by cloud, we ourselves may be the hottest body in the surroundings. Then *we* tend to radiate heat to our surroundings, for example to a snow-covered ground which is at a much lower temperature. (In winter, of course, we lose heat also by conduction and convection in the surrounding air, but these are separate processes and do not influence the radiation.) In the same way, an X-ray tube keeps cool partly by radiating heat to its cooler surroundings (11.3.4).

When radiation of any kind falls on a body, the body is said to be **irradiated** and the energy may be **absorbed; absorption** is the reverse process to radiation or **emission**. In general, the ease with which *heat* is emitted or absorbed by a body depends on the nature of its surface, as well as on its temperature. Black or matt objects both emit and absorb heat more effectively than white or polished objects. For this reason we wear white in summer so as not to *absorb* too much heat from the sun; in winter we should also wear white so as not to *radiate* (emit) too much heat to our surroundings. For the same reason, the casing in which X-ray tubes are mounted is often painted black so that the tube will remain as cool as possible.

2 Atomic structure

2.1 THE STRUCTURE OF THE ATOM

2.1.1 Elements and compounds. All matter is made up of chemical substances which can be divided into two kinds, elements and compounds.

DEFINITION **An element is a distinct kind of matter which cannot be decomposed into two or more simpler kinds of matter.**

DEFINITION **A compound is formed when two or more elements combine together chemically to produce a more complex kind of matter.**

For example, hydrogen and oxygen are both *elements;* neither can be decomposed into simpler kinds of matter. They can combine together to make a more complex kind of matter, viz. water, which is a *compound* of hydrogen and oxygen.

2.1.2 Atoms and molecules. A sample of an element is made up of extremely small entities called **atoms.**

DEFINITION **Atoms are the smallest particles of an element that can exist without losing the chemical properties of the element.**

DEFINITION **Molecules are the smallest particles of a compound that can exist without losing the chemical properties of the compound.**

Molecules are combinations of atoms, for example, a molecule of water is composed of two atoms of hydrogen and one of oxygen; this fact underlies the laws of chemical combination, in which elements combine together in simple proportions.

The diameter of an atom is of the order of 10^{-10*} metres, which is too small to be seen even under the most powerful microscope; the central core of an atom, called the **nucleus,** has a diameter only one ten-thousandth of this. It is difficult to form a mental picture of such small sizes; about ten million million atoms would be required to cover one full-stop.

2.1.3 Protons, neutrons and electrons. Our present concept of the structure of the atom is based on the work of Rutherford and Bohr early in the twentieth century. This provides a simple picture of the atom as an electrical structure having three basic units, the proton, the neutron and the electron, and it is adequate for the discussion of most phenomena in radiological physics.

The simplest atom is one of the element hydrogen. This consists of a central nucleus comprising one proton around which one electron moves in a **shell** or **orbit.**

The **proton** (Table 2.1) is a *heavy* particle carrying a *positive* electric charge (3.1.1). The **electron** is a very much lighter particle having a mass of

TABLE 2.1 (2.1.3) *The atomic particles*

	Symbol	Electric charge (atomic units)*	Mass (atomic mass units)†	Found in:
Proton	p	+ 1	1·00728	Nucleus
Neutron	n	0	1·00867	Nucleus, except $^1_1$H
Electron	e	−1	$\frac{1}{1840}$	Shells around nucleus

*The charge on the proton is numerically equal (but of opposite sign) to that on the electron and is equal to $1·602 \times 10^{-19}$ coulomb (3.3.1).

†The atomic mass unit is defined as $\frac{1}{12}$ of the mass of the most abundant isotope of carbon and is equal to $1·66 \times 10^{-27}$ kilogram.

only $\frac{1}{1840}$ of the mass of the proton, and it has a *negative* charge of exactly equal magnitude (but of opposite sign) to that of the proton. Consequently almost all the mass of the atom is in the nucleus, and the positive charge of the nucleus is balanced by the negative charge of the electron to make the atom as a whole electrically neutral.

$* \; 10^{-10} = \dfrac{1}{10\,000\,000\,000}$

The next simplest atom is one of the element helium. In this, the nucleus comprises two protons and two neutrons and there are two orbital electrons moving around the nucleus. A **neutron** (Table 2.1) is a particle with a mass approximately equal to that of a proton but with *no* electric charge. The nucleus of the helium atom therefore has a positive charge of 2 units (due to the two protons) and a mass of approximately 4 units. Also, the two positive charges of the nucleus are balanced by the two negatively charged electrons around the nucleus.

2.1.4 Atomic number and mass number. An element can be identified either by its **atomic number** (Z) or by its name.

DEFINITION **The atomic number of an element is the number of protons in the nucleus (which is equal to the number of electrons around the nucleus) of an atom of that element.**

In going from hydrogen to helium, for example, the number of protons in the nucleus (and also the number of orbital electrons) has increased from 1 to 2; thus the atomic number of hydrogen is 1 and that of helium is 2.

The atomic number of an element determines its chemical properties because these depend on the number of electrons present and on the way in which they are arranged around the nucleus (2.2.2).

DEFINITION **The mass number of an atom is the total number of protons and neutrons in the nucleus.**

The mass number gives a measure of the mass of the nucleus. Protons and neutrons are known collectively as **nucleons** because they are found in the nucleus.

When it is necessary to indicate the values of the atomic number and the mass number, they are often added to the chemical symbol for the element as prefixes, the mass number being placed above the atomic number. Thus the helium described above is represented as $^4_2$He. This is an example of a **nuclide**.

DEFINITION **A nuclide is a particular variety of atom characterized by a given atomic number and a given mass number.**

2.2 THE PERIODIC TABLE

2.2.1 The arrangement of the elements. The elements can be arranged in order of increasing atomic number to form the **periodic table** (Table 2.2) of

TABLE 2.2 (2.2.1) *Examples from the periodic table**

Atomic number	Name of element	Chemical symbol	Mass numbers of isotopes‖	Average atomic weight†	Comment
1	Hydrogen	H	1	1·008	Lightest element‡
2	Helium	He	4	4·003	Nucleus is an alpha particle (16.2)
3	Lithium	Li	7	6·94	
6	Carbon	C	12	12·01	‡
7	Nitrogen	N	14	14·01	
8	Oxygen	O	16	15·999	
13	Aluminium	Al	27	26·98	Used for X-ray filters (14.5) and step-wedges
17	Chlorine	Cl	35, 37	35·5	Example of isotopes (2.3)‡
19	Potassium	K	39	39·1	‡
20	Calcium	Ca	40	40·1	Bone mineral‡
27	Cobalt	Co	59	58·9	‡
29	Copper	Cu	63, 65	63·5	Used for X-ray filters (14.5)
50	Tin	Sn	116, 118, 120	118·7	Used for X-ray filters (14.5)
53	Iodine	I	127	126·9	X-ray contrast medium‡
55	Caesium	Cs	133	132·9	$^{137}_{55}$Cs in fall-out from nuclear weapons‡
56	Barium	Ba	137, 138	137·3	X-ray contrast medium
74	Tungsten	W	182, 183, 184, 186	183·9	Used for targets in X-ray tubes
79	Gold	Au	197	197·0	‡
80	Mercury	Hg	198, 199, 200, 201, 202	200·6	‡
82	Lead	Pb	206, 207 208	207·2	Used for X- and gamma-ray shields
86	Radon	Rn §	(222)	(222)	Used in radiotherapy
88	Radium	Ra §	(224), (226), (228)	(226)	Used in radiotherapy
92	Uranium	U §	238	238·0	Heaviest naturally occurring element

*The name periodic table arises because elements with similar arrangements of their atomic electrons and therefore with similar chemical properties occur periodically throughout the Table (2.2.2).

† As determined in a chemical experiment in which the sample of the element would contain a very large number (greater than 10^{20}) of atoms of the naturally occurring isotopes. The small contribution from the mass of the electrons would be included.

‡ Man-made radioactive isotopes of these elements are used in medicine.

§ Occur in nature as members of radioactive series (16.1.3).

‖ Mass numbers of most abundant (>10%) naturally occurring isotopes.

17

the elements. Hydrogen and helium are the first and second elements in the table and the third element is called lithium. The nucleus of an atom of lithium therefore contains three protons. It also contains four neutrons, giving a mass number of 7, and is surrounded by three orbital electrons (Fig. 2.1). Note that two of the orbital electrons are accommodated in one shell, the K-shell, while the third electron is in another shell, the L-shell, farther out from the nucleus.

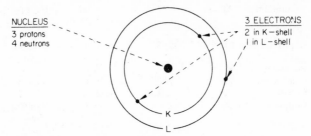

Fig. 2.1   An atom of lithium, $_3^7$Li (not drawn to scale).

2.2.2   Electron shells. There are natural laws governing the way in which electrons are 'accommodated' around the nucleus. For example, in an atom of aluminium. $_{13}^{27}$Al, there are thirteen orbital electrons arranged as illustrated in Fig. 2.2. The K-shell contains two electrons, the L-shell eight electrons, and the M-shell three electrons. In more complex atoms with more electrons to accommodate, there are yet more shells taking successive letters of the alphabet the farther out they are from the nucleus.

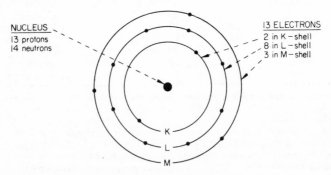

Fig. 2.2   An atom of aluminium, $_{13}^{27}$Al.

The number of electrons in the outermost shell of an atom largely governs the chemical properties of the atom (2.1.4). In going through the elements of the periodic table, there is a repetitive pattern in the way in which the outer electrons are arranged in their atoms and consequently there is a repetition of chemical properties — hence the name, periodic table. For example, in helium, neon, argon, krypton and xenon, the outermost shell has a full complement of electrons in each case and the chemical properties of these elements are similar — they are all chemically inert gases.

2.3 ISOTOPES

If there were only one kind of atom for each element, we would expect the average weight of the atoms of each element to be a whole number (neglecting the relatively small mass of the electrons in an atom and taking the masses of the proton and the neutron each to be 1 (Table 2.1)). In practice, however, the average atomic weight for a sample of an element can be far from a whole number. In the case of chlorine, for example, the average atomic weight of a sample is 35·5 (Table 2.2). This is because two kinds of chlorine atom are present. All the atoms in the sample must have an atomic number of 17 (or they would not be chlorine) but 75% of them have a mass number of 35 and the other 25% a mass number of 37, resulting in an average atomic weight of 35·5. The difference in mass numbers arises because the lighter atoms have eighteen neutrons in the nucleus and the heavier ones have twenty. These two kinds of chlorine are known as **isotopes** of chlorine.

DEFINITION   **Isotopes are nuclides which have the same number of protons but different numbers of neutrons in their nuclei.**

Isotopes therefore have the same atomic numbers as each other (they are the same element) but have different mass numbers.

There are natural laws which relate to the proportion of neutrons to protons in the nucleus of an atom. The nuclei of the two isotopes considered above, $^{35}_{17}Cl$ and $^{37}_{17}Cl$, are both *stable,* but the nucleus of another isotope of chlorine, $^{36}_{17}Cl$, which has a different proportion of neutrons to protons, is *unstable* and undergoes **radioactive disintegration** (16.1.1). $^{36}_{17}Cl$ is therefore a **radioactive isotope** or **radioisotope** of chlorine.

2.4 IONIZATION AND EXCITATION

An atom taken as a whole is normally electrically neutral (2.1.3). If, however, one or more of the orbital electrons are removed from the atom, the remainder of the atom is left positively charged and is known as a positive ion. This process of removal of orbital electrons is known as **ionization**.

To produce ionization, energy must be given to an orbital electron (which is negatively charged) in order to remove it from the atom against the force of attraction (3.1.2) which binds it to the positively charged nucleus. The energy which is just sufficient to do this is known as the **binding energy**. The magnitude of the binding energy depends on the atomic number (i.e. on the number of positive charges on the nucleus) and on the shell from which the electron is being removed. It is greater for elements of higher atomic number and also greatest for the K-shell (the shell nearest to the nucleus).

Under some circumstances, it is possible for one or more additional electrons to attach themselves to a neutral atom producing a net negative charge. In this way, a *negative* ion is formed.

The energy required to produce ionization may be supplied to the atom by several means of which an important one is **ionizing radiation**. This is the general name given to X rays (Chapter 10), gamma rays, alpha particles and beta particles (Chapter 16), etc.

Sometimes the energy given to an atom is less than the binding energy and an orbital electron receives insufficient energy to enable it to leave the atom as in ionization. Instead, the electron may move from its original shell, either to another shell farther out from the nucleus or to some other **energy-level** characteristic of the particular atom. The atom then has more energy than in its normal state. It is said to be in an **excited state** and the process is known as **excitation**. The atom does not remain in an excited state for long, the vacancy in the original shell being filled by an electron moving in from the outer shell or energy-level with the emission of energy equal to the excess.

3 Electricity

(In this chapter the term 'electric' will be used in place of the more traditional 'electrostatic'. Besides being increasingly common usage, we prefer this form because electric fields may often be changing rapidly, i.e. they may be far from 'static'.)

3.1 ELECTRIC CHARGES

3.1.1 Frictional electricity. In Chapter 2 we explained how atoms are made up of sub-atomic particles, of which electrons and protons were said to be **electrically charged**. The meaning of this expression was not then defined but, although we cannot *explain* the phenomenon of **electricity**, we can convey an idea of its nature by describing some of its properties.

In section 2.4 we described how a negative electron, which is very light, can be removed from an atom leaving a positive, heavy remainder. This process is called **ionization**. A very simple but crude way of separating electrons from atoms for demonstration purposes is by rubbing two materials together, i.e. by friction. The electric charges so produced are called **frictional charges.**

This effect can be very easily demonstrated by rubbing a piece of polythene rod with some dry nylon. It will be found that both the rod and the nylon are then capable of attracting small pieces of light material like expanded polystyrene (such as may be broken from an ordinary white 'ceiling tile'). The explanation of this attraction will be given later, but for the present it suggests that separate electric charges have been produced by friction on the polythene and on the nylon; the former is in fact negative and the latter positive. Hence electrons must have been rubbed off the nylon and on to the polythene, leaving positively charged ions behind on the nylon. In solids it is always the negatively charged electrons that move. Movement of the positive ions (which contain the atomic nuclei) would imply movement of the material itself, which does not of course happen.

If the same experiment is tried with a rod made of metal instead of polythene, no frictional charges can be observed. This is because the metal allows the frictional charges to flow through it easily, through the experimenter's hand and down to the earth. (The earth acts as a large reservoir of charge.) Substances which allow electrons (electricity) to flow through them easily are called **conductors** (compare **heat conduction**, 1.4.3). Carbon and most metals are conductors, one of the best being copper. Substances which *prevent* the flow of electrons are called **insulators**; examples are plastics, rubber and glass. Electrical conduction will be further explained in section 4.1.1.

3.1.2 The laws of electric force; electric fields. Some properties of the electric charges produced by friction can be demonstrated by a simple experiment using two small balls of expanded polystyrene that have been suspended about 20 mm apart by light, insulating threads of silk or nylon (Fig. 3.1a). (In practice, the polystyrene balls should first be coated with Indian ink (carbon) to make their surface electrically conducting.) If both balls are charged similarly by bringing them into contact with the polythene, i.e. negatively, or with the nylon, i.e. positively, the balls will move away from each other (Fig. 3.1b and c). If however one ball is charged positively and the other negatively, the two will move together (Fig. 3.1d). These movements are caused by forces of **repulsion** and **attraction** respectively. In fact, a force is exerted between any two electric charges, and it will cause movement if the charges are free to move. The repulsion and attraction of two electric charges are summarized in the **law of electric force**:

DEFINITION   **Similar charges repel, dissimilar charges attract.**

The force of attraction can be compared with the force of **gravitation** (1.3.3), although the two are of quite different natures. Like gravitation, the force *decreases* rapidly as the distance between the two bodies *increases*. If the charges are small in extent compared with the distance between them, the law of attraction (and repulsion) obeys the well-known **inverse-square law**:

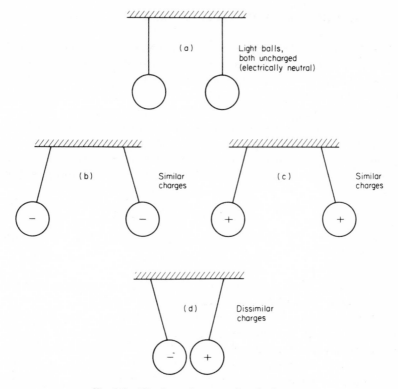

Fig. 3.1   The forces between electric charges.

DEFINITION   **The force between two charges is inversely proportional to the square of the distance between them.**

Thus if the distance is doubled, the force is reduced to one-quarter; if the distance is trebled, to one-ninth, etc. If the distance is very large indeed (infinite), the force will be negligible.

There is another way of thinking about electric and gravitational forces which often makes problems clearer. We say that electric charges (for example) are surrounded by **electric fields**; a **field** is simply a region in which electric force is exerted on another charge, which is itself surrounded by an electric field. Alternatively, we can say that the two electric fields *interact*, with the result that a force of attraction (or repulsion) is produced.

23

3.2   ELECTRIC INDUCTION; ELECTROSCOPES

3.2.1   Electroscopes and electrometers. The experiment with two suspended polystyrene balls (3.1.2) illustrates the principle of the **electroscope**, which is a useful laboratory instrument for indicating the presence of an electric charge and its sign. Fig. 3.2a shows the basic construction. A metal rod, with a disk at its upper end, passes through a stopper made of a very good insulator (such as polythene) into a metal box with glass front and back. The lower end of the rod has fixed to it a piece of gold leaf, which is very light and which therefore requires very little force to raise it. This instrument can be used to indicate the presence of an electric charge, in the following way.

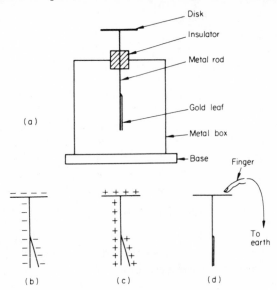

Fig. 3.2   The electroscope.

If a negative frictional charge is transferred to it by contact with the polythene in section 3.1.1, the electrons distribute themselves all over the rod and the gold leaf. Because similar charges repel, the leaf is repelled from the rod and rises (Fig. 3.2b). Exactly the same behaviour is observed if electrons are *removed* from the electroscope, by bringing it in contact with the charged nylon (Fig. 3.2c).

To cause the leaf to fall, i.e. to **discharge** the electroscope, all we need to do is to touch the metal disk with our finger, when the charge will flow away through our body to earth, leaving the electroscope uncharged (Fig. 3.2d).

Common sense tells us that the distance the gold leaf rises depends on the size of the charge; a large charge, positive or negative, produces a large force of repulsion, and therefore a large movement, and vice versa. The electroscope described above is a relatively crude instrument, but there exist more refined electroscopes called **electrometers** which can be used to measure charge very accurately. A great deal of the early fundamental research work in physics was carried out with relatively simple apparatus such as electrometers, long before the age of highly complex electronic apparatus, computers, etc.

3.2.2 Electric induction. In section 3.1.1 we described how a body charged with frictional electricity can attract small, light bodies such as pieces of expanded polystyrene. This effect may at first seem quite natural, but on second thoughts how is it possible for a *charged* body to attract an *uncharged* body? The answer lies in the phenomenon called **electric induction.**

Figure 3.3a shows a horizontal cylindrical conductor made of metal, with rounded ends, mounted on an insulating stand so that charge cannot leak away from it. Initially the whole conductor is electrically neutral. Suppose now we bring near one end of the conductor a negatively charged body. This negative charge will repel electrons in the conductor, making the far end negative and leaving the near end positive as in Fig. 3.3b. (Remember that in solids it is only the negative charges, i.e. the electrons, that move.) A positively charged body will have the opposite effect (Fig. 3.3c). This process is called **electric induction.** We shall meet other kinds of **induction** in this book; the word generally implies some kind of action at a distance.

We can now see how an initially *uncharged* body can be attracted. In Fig. 3.3b and c, the conducting cylinder will experience an attracting force from the charged body. Although the cylinder is electrically neutral *as a whole,* some of its charges have been separated in space, and the *dissimilar* charges are closer together than the *similar* charges. Hence the force of attraction between the former is greater than the force of repulsion between the latter. Therefore the net force is one of attraction.

Another example of the application of electric induction is to enable us to determine the **polarity** of the charge (i.e. whether it is positive or negative) on an electroscope. Both positive and negative charges produce the same type of

25

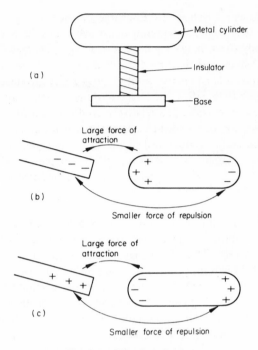

Fig. 3.3 Electric induction.

deflexion, so how can we tell which is which? Suppose the electroscope is negatively charged, and the gold leaf is deflected moderately. If then we bring a negatively charged body *near* to the disk of the electroscope (but not touching it) more negative charge will be repelled *down* the electroscope rod (by electric induction) and the leaf will rise further. If on the other hand we bring a positive charge near, electrons will be attracted *up* the rod and the leaf will fall.

3.3 ELECTRIC CHARGE AND ELECTRICAL POTENTIAL

3.3.1 Units of electric charge. Electric charges result from the separation of negative electrons from atoms, leaving behind positive ions. It would thus be natural to measure electric charge simply in terms of the *number of electrons,* either as an *excess* for a *negative* charge or as a *deficit* for a *positive* charge.

However, this would be very inconvenient in practice, because the electron has such a very small charge that very large numbers would have to be used. It would be like trying to buy a quantity of, say, petrol in terms of a number of molecules instead of gallons or litres!

Hence there is an S.I. unit of electric charge, formally defined in terms of length, mass and time, called the **coulomb**. This is a *derived* unit (1.2.3), and the way in which it is derived from the *basic* units is too complex to be discussed here. We shall therefore state simply:

DEFINITION   **The coulomb is a measure of electric charge or quantity, equivalent to that on about 6 × 10^{18} electrons.**

The reader is reminded that 6 × 10^{18} is a very large number indeed, viz. six million million million, yet the coulomb is not such a very large unit of charge. It is the amount of electricity which would flow in a pocket torch in about five seconds! This example emphasizes just how small the charge on an electron really is.

Although the coulomb is widely used as a practical unit of charge there are other units. One of these, called the **electrostatic unit of charge (e.s.u.)**, must be mentioned and defined because it is the basis of the former definition of the roentgen (15.2.1). As with the coulomb, the e.s.u. of charge has a formal definition, but here we shall say simply:

DEFINITION   **1 e.s.u. of charge is equivalent to that on about 2 × 10^9 electrons.**

From the last two definitions it can be calculated that

$$1 \text{ coulomb} = \text{about } 3 \times 10^9 \text{ e.s.u.}$$

3.3.2   The behaviour of electric charges. We have already described a number of properties of electric charges, such as their ability to flow through **conductors**, and their attractive or repulsive force on each other. We must now discuss these properties further, with the object of emphasizing certain practical applications.

Let us consider the cylinder of conducting material illustrated in Fig. 3.3a (3.2.2). A charge, i.e. a quantity of electrons, placed on one end will distribute itself (not necessarily uniformly, see below) all over the conductor. However, it will be confined to the *surface* of the conductor, because the electrons tend to repel each other and therefore to drive each other outwards on to the surface. If instead the cylinder is made of an *insulating* material, the electrons (i.e. the charge) will remain where they are put. Often an insulator will appear to behave as a conductor, however, because of a layer of moisture which may have condensed on the surface of the insulator. Moisture, particularly if it contains dissolved salts, is a fairly good conductor; thus it is *very dangerous* to operate electrical apparatus with wet hands (making it easier for charge to flow through the body to earth, section 3.2.1) unless the apparatus is especially designed for the protection of the operator (12.3.2).

The distribution of charge on a conductor depends entirely on its shape, if there are no other charged conductors nearby. Fig. 3.4 shows three shapes of conductor, each on an insulating stand, and each charged with the same number of electrons (i.e. each carrying the same negative charge in coulombs). The way in which the *density* of the charge varies over the three shapes, viz. a sphere, a cylinder and a pointed pear-shaped solid, is shown qualitatively by the density of the negative signs on the diagram.

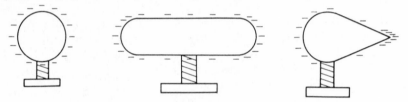

Fig. 3.4   The distribution of charge on conductors.

It is clear that the density of charge is larger at curved surfaces, and is particularly large at the point of the pear-shaped solid. Because of this very large concentration of charge at points, we say that the value of the electric field (3.1.2) in the neighbourhood of the point is very large. The electric field, if of large enough value, may cause ionization of the atoms in the air; the positive ions thus formed collect electrons from the point and the conductor loses charge into the air. This is called an **electric discharge**; it must be avoided in the design of certain parts of X-ray apparatus where such large electric fields are often produced. This is done by the very simple precaution of avoiding sharp edges and points in conductors; thus conductors in X-ray work are usually obviously rounded in shape and are often polished so that they have as few sharp irregularities on their surfaces as possible.

3.3.3   Electrical potential. In section 1.3.5, we explained that water flows from a *higher* level to a *lower* level. This is because its **potential energy** is greater at the higher level than at the lower level. In general, matter tends to move so as to reduce its energy; in so doing it converts the energy into work or into other forms of energy. For example, falling water can be used to turn a mill-wheel which can do useful work.

In this section we shall see that electricity shows a precisely similar behaviour and that this is linked closely with the idea of potential energy; the idea leads to the very important concept of **electrical potential.**

Let us consider a body having a negative charge of Q coulombs (Fig. 3.5).

Fig. 3.5   Electrical potential.

We shall disregard the mass of the body so that no gravitational forces will complicate the discussion (1.3.3). Suppose another similar body, having a negative charge of 1 coulomb, is situated at **infinity** – that is, at such a distance that the repulsive force between them can be neglected (inverse-square law, 3.1.2). Now let us bring the unit charge nearer Q; as it comes nearer we must exert more and more force on it until it reaches, say, the point A. We have had to do mechanical work on the unit charge to bring it to A; that work has been converted into **potential energy,** just as work done on the mass in Fig. 1.1 (1.3.4) was converted into potential energy at position 1. Moreover, like the mass, if we release the unit charge at A, it will be repelled back to infinity by Q and, in the process, it will in principle be able to do work, although in practice it would be rather difficult to devise a suitable arrangement. Similarly, the unit charge would have different amounts of potential energy at other points B, C, D, etc., situated around the charge Q. We say that each point A, B, C, D has an **electrical potential** associated with it, and this is defined in terms of the work done in bringing unit charge to the point. The S.I. unit of potential is the **volt (V).**

DEFINITION   **The electrical potential in volts at a point is equal to the work done in joules in bringing one coulomb of positive charge from infinity to the point.**

Two points of explanation are necessary in this definition. First, the specification of a *positive* unit charge instead of the *negative* unit charge used in the example above is a survival of the time, before electrons were discovered, when electricity was thought to consist of two 'fluids', positive and negative (4.1.3). It is unfortunate that, by chance, definitions were then couched in terms of the *positive* 'fluid', whereas now we know that the *negative* 'fluid' is the one which commonly moves in solids. The above definition thus relates to *positive* electrical potential, but it can also be stated quite correctly in the negative form:

DEFINITION   **The negative electrical potential in volts at a point is equal to the work done in bringing one coulomb of negative charge from infinity to the point.**

The second point to be explained might help to clarify the first. In Fig. 3.5, infinity is chosen as the absolute **zero** of electrical potential energy (1.3.5). This is analogous to the use of sea level as the zero of height in Fig. 1.1. But of course *negative* heights 'above' sea level are possible; these of course correspond to depths *below* sea level. Hence we can equally well refer to positive *and* negative electrical potential so long as we clearly say which is meant at the time. In this book, we shall write mostly of negative electricity because it is electrons that move, always in solids and often in space; we shall therefore refer most often to negative potential.

The idea of **potential** at a point is so important that before proceeding to the next section we must repeat the important principle that it is the electrical potential at a point that determines the direction in which electric charge will flow. In particular, negative charge (electrons) will flow from a point of high negative potential to a point of low negative potential. This behaviour is analogous to the flow of water from a greater to a lesser height. The idea of charge flowing from high to low potential emphasizes the importance of the concept of *difference* of potential or **potential difference** (p.d.); this will be elaborated in section 4.1.2, etc. Although, from the above definitions, it is apparent that the absolute zero of potential is situated theoretically at infinity, this is not a very useful concept in practice. In the everyday use of electricity, therefore, we regard **earth** as our **relative zero** of potential, because it is so large that its potential may be regarded as virtually constant (3.1.1), just as sea-level is unaltered by pouring in a small quantity of water.

Potentials encountered in practice have very widely differing values, so that the volt is sometimes too large, sometimes too small, a unit. In radiology, examples of useful multiples and sub-multiples are the **kilovolt** (kV) ($1 \text{ kV} = 10^3 \text{ V}$) and the **millivolt** (mV) ($1 \text{ mV} = 10^{-3} \text{ V}$).

3.4 CAPACITANCE AND CAPACITORS

3.4.1 Capacitance. Water has very many uses; we therefore need to be able to store it temporarily in various types of vessel: drinking glasses, bottles, tanks, reservoirs, etc. Similarly it is very convenient to be able to store electricity (electric charge) for short periods of seconds or minutes; this may be done in conductors or in special arrangements of conductors called capacitors (formerly called condensers). (Longer periods of storage demand different methods which will be mentioned in section 5.3.1.)

We know that we can store electrons on a conductor (e.g. Fig. 3.4, 3.3.2); what determines the quantity of electricity (in coulombs) that can be stored in a given conductor? For example, we can speak of the *capacity* of a water tank; this means the number of gallons of water it will hold without overflowing. Can we apply a similar criterion to the storage of electric charge? It all depends on what we mean by *overflowing* in the electrical case; this could only refer to the loss of charge via ionized air (3.3.2) and is not such a well-defined or constant criterion as the overflowing of a tank. For this and other reasons it is usual to refer to the **capacitance** (*C*) (the present-day word for electrical capacity) in a different way.

This can best be understood by again considering a water analogy. Fig. 3.6a shows two cylindrical vessels, both of which have been filled to unit depth (i.e. 1 metre) with water. Obviously the wide vessel of base area A_2 (m^2) will require a greater quantity of water (Q_2) than the narrow vessel A_1(m^2) which requires a quantity of only Q_1 of water. Fig. 3.6b shows the electrical equivalent. Here we have a small conductor (a metal sphere) and a large one. If we 'pour' negative charge into each conductor we shall find that their negative potential (*V*) will rise (3.3.3), and our water analogy suggests that the smaller conductor will need a smaller charge (Q_1) to raise the negative potential by 1 V than will the larger conductor (Q_2). We therefore define the capacitance of a conductor in terms of the charge required, not to make it 'overflow', but to raise its potential (electrical level) by unit amount. The S.I. unit of capacitance is the **farad** (F).

DEFINITION **The capacitance of a conductor in farads is equal to the electric charge in coulombs required to change the potential of the conductor by 1 V.**

The capacitance of a conductor is thus analogous to the base area of a vessel in Fig. 3.6a.

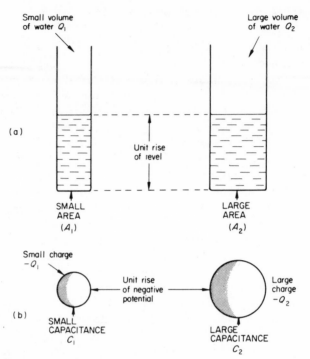

Fig. 3.6   Electrical capacitance.

From this definition we can derive an important relation or 'formula': if it requires a charge numerically equal to C coulombs to change the potential of a conductor of capacitance C farads by *one* volt, then it must require a charge (Q) equal to CV coulombs to change its potential by V volts.

Hence we can say

$$Q = CV, \qquad \left(C = \frac{Q}{V} \quad \text{or} \quad V = \frac{Q}{C} \right). \qquad \text{Eq. 3.1}$$

This is an important equation which must be learnt by heart (and understood!).

3.4.2   Capacitors (formerly called condensers). The **farad** (F) is too large a unit for normal use, and in practice capacitancés are usually expressed in microfarads (μF^*) $(1 \ \mu F = 10^{-6} \ F)$. However, the capacitances of isolated

$*\mu$ is the Greek small letter 'mu'.

conductors are very small indeed and quite inadequate for most practical purposes. For example, the capacitance of a sphere of radius 10 mm is only about $10^{-6}\,\mu F$. Some means of increasing the effective capacitance of conductors must therefore be found; this may be done by using special arrangements of conductors called **capacitors** (condensers). (The older name arose from the idea popularly believed at one time that the electricity was in some way being *condensed* into a smaller space; hence the capacitance of a given conductor would be larger.)

Figure 3.7 illustrates the principle of the capacitor. Fig. 3.7a shows an isolated conductor in the form of a flat conducting plate to which connexion can be made by a wire. A negative charge is put on the plate, resulting in a rise of negative potential. The capacitance is very small.

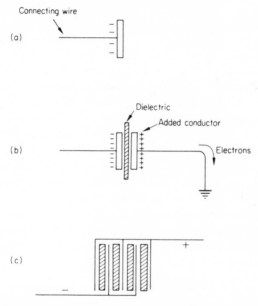

Fig. 3.7   The principle of the capacitor. (a) Isolated conductor: small capacitance; (b) capacitor: large capacitance; and (c) a multi-plate capacitor with four dielectrics.

In Fig. 3.7b a similar plate has been brought up very close to the first, but separated from it by a thin insulating layer called the **dielectric**. The second conducting plate is charged positively, either deliberately, or automatically by **electric induction**. The latter operates as follows. The second plate is connected to earth, which acts as a reservoir of charge (3.1.1); when the second plate is brought near the first plate, the negative charge on the first plate repels some of the negative charge on the second plate and causes it to flow to earth, leaving behind an equal positive charge. But this positive charge is very near to the negative charge on the first plate, and has the effect of reducing its negative potential. This can easily be seen by qualitatively comparing the work necessary to bring unit negative charge up to the first plate; it will be much less in Fig. 3.7b than in Fig. 3.7a because of the attracting force of the new positive charge. Hence for the same charge Q, the potential V has fallen. From Eq. 3.1 (3.4.1), $C = Q/V$, and as V is less and Q constant, C must be greater. Hence we have succeeded in increasing the capacitance of the original conductor, perhaps by many thousands or millions of times.

Equation 3.1 (3.4.1) can be applied also to capacitors, but the value of Q to be used is the charge on *one* of the plates, not on both, and V is the **potential difference** between the plates. The capacitance of a capacitor depends on several factors. Often capacitors are made with many plates, arranged in two sets interleaved with insulating **dielectrics** (Fig. 3.7c). Clearly the capacitance must increase with the number of plates and with their area, and also with decrease of the distance between adjacent plates (because the effect of one charge on the other is thereby increased). The nature of the dielectric also has an effect; a capacitor with a vacuum between the plates has the least capacitance, and any other material will increase the capacitance above the value for a vacuum by a factor called the **dielectric constant** (K) of the material.

If A is the area of overlap of adjacent plates,
  n is the number of dielectrics,
  d is the distance between adjacent plates, and
  K is the dielectric constant,

$$C = \frac{8 \cdot 85 \, nKA}{d} \times 10^{-12} \text{ F.}$$ Eq. 3.2

Capacitors are used for many purposes in radiology, to be explained later; details of their construction may be obtained from reference books. However, there is an aspect of their construction other than their capacitance which is of considerable practical importance, viz. their **breakdown voltage**. The term **voltage** we shall explain later (5.3.2); here it is synonymous with potential difference. A capacitor breaks down at a potential difference between its plates when the electric field so produced is too great for the dielectric to withstand. The material of the dielectric punctures, a spark passes and the capacitor is *discharged*. This is the electrical condition which is analogous to the overflowing of a water tank (3.4.1). For reliability, a capacitor must always be used at a potential difference well below its breakdown voltage.

The ability of an insulating material to withstand electric fields, whether it forms the dielectric of a capacitor or is simply used to insulate one conductor from another, is represented by a quantity called its **dielectric strength**. Air has a low value of dielectric strength, whereas oil and rubber have high values. Therefore oil and rubber are widely used in X-ray generators where large potential differences are found.

4 Electric current

4.1 ELECTRIC CURRENTS

4.1.1 Electron flow. In Chapter 3 we discussed the 'production' (or rather separation) of electric charges; this was said to result from the **movement of electrons.** However, we were then interested mainly in the properties of the stationary or **static** charges so produced. In this chapter we shall discuss the properties of the electron movement itself; this movement is called an **electric current** and is of very great importance in the world of today.

In section 3.1.1 we introduced also the idea of **conductors** and **insulators.** The movement of electrons, i.e. an electric current, will take place only through conductors; this is thought to be due to the presence in conducting materials of **free electrons** which, as their name implies, are free to move with little hindrance in the spaces between the atoms of the conductor. Insulators are believed not to have these free electrons; thus electron flow through them is not possible.

4.1.2 A simple electrical circuit. In section 3.3.3 it was emphasized that electrons flow only as a result of a **difference of potential,** i.e. from a point of high negative potential to a point of low negative potential. Hence to maintain a *constant* flow of electrons in a conductor there must be some means of maintaining a *constant* potential difference between the ends of the conductor. Fig. 4.1a shows one way in which this may be simply achieved for demonstration purposes. A conductor in the form of a length of fine copper wire is **connected** (i.e. electrically joined) between a copper (Cu) and a zinc

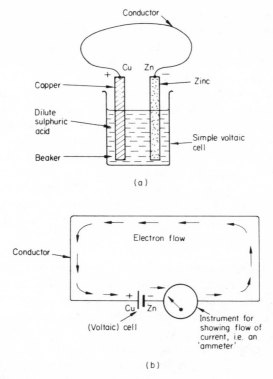

Fig. 4.1 A simple electrical circuit.

(Zn) plate immersed in dilute sulphuric acid in a glass jar. The two plates and the acid together are called a *simple voltaic cell,* usually abbreviated to **simple cell** or just '**cell**'. Because of chemical reactions occurring between the plates and the acid, the cell has the property of maintaining a potential difference or p.d. between the plates (points Cu and Zn in Fig. 4.1a). The point Zn is found to have a higher negative potential than point Cu, by about one volt, and this causes a continuous flow of electrons, i.e. an electric current, through the conductor.

How do we know that this current is flowing? It can be detected in a number of ways which will be discussed later (5.1.5). However, for the present let us assume the existence of an instrument for indicating the flow of current, called an **ammeter**. Fig. 4.1b shows the arrangement of Fig. 4.1a (plus the ammeter) in the form of **conventional symbols**. The whole forms a simple **electrical circuit.** When drawing **circuits,** most of which are *much* more complex than Fig. 4.1b (e.g. that of an X-ray generator!), it would be impossible to make an artist's drawing of each component every time. So all components are represented by standard conventional symbols as shown. For example, the cell symbol consists of a short thick line and a longer thin one; the former is assumed to be at the higher negative potential (it is often marked −) and the latter at the higher positive potential (often marked +). Electrons will therefore flow from the *negative* (black) **terminal** (a terminal is simply a point at which a connexion is made) through the conductor and back to the *positive* (red) terminal. For a current to flow it is essential to have a complete return path; the electrons must have a point of lower negative potential to return to, i.e. the positive terminal. The ammeter is connected *in* the circuit, as shown, because for the flow of electrons to be indicated they must go *through* the ammeter. The ammeter is said to be **in series** with the circuit.

4.1.3 Conventional current and electromotive force. We have said many times that we know that only *negative* charges (electrons) move in solids. Movement of *positive* charges would imply that the copper ions themselves were migrating along the conductor; in other words, the material of the wire itself would be 'crawling' along, which of course does not happen. Nevertheless, before the existence of electrons was realized, physicists knew that 'something' was flowing in the wire, and guessed (wrongly) that it flowed from the positive to the negative terminal. Many rules were formulated in terms of this imaginary flow, which is now known as **conventional current.** We shall be compelled to present these rules in subsequent chapters (6.2.2, 7.1.3), and will therefore sometimes talk about conventional current, which can be thought of as flowing in a direction opposite to the *real electron* flow. It may be that in the foreseeable future the rules will be altered by international agreement, but this unfortunately has not yet happened.

The electron flow or current in Fig. 4.1 (4.1.2) will go on flowing so long as the cell is able to maintain a p.d. between its terminals. This it is able to do by virtue of the chemical processes which go on within it. The result is that a negative *electric charge* flows 'downhill' through a *difference of potential,* and, like a mass falling 'downhill' through a *difference of height* (1.3.5), this flow is associated with a certain amount of *energy,* in this case *electrical* rather than *mechanical.* The **electrical energy** cannot be created from nothing, of course (1.1.4); it is in fact converted from **chemical energy** in the cell. The ability of the cell to maintain a potential difference between its terminals as a result of converting *chemical* into *electrical* energy is called its **electromotive force** or **e.m.f.** (see section 5.3.2 for further explanation).

4.1.4 The unit of current is called the **ampere (A).** Before defining the ampere it is as well to understand clearly what we mean by 'amount of current'. Suppose we could be endowed with superhuman powers and could look into the current-carrying conductor at a single point with a stop-watch in our hand. If we could then count the number of electrons passing this point in one second, this could serve as a measure of current: so many electrons per second. However, this would have the disadvantage we have met previously of numbers of very great size; instead, let us imagine that we can count the number of **coulombs per second.** This would be equal to the amount of current in amperes. In a similar way the current of water in a pipe can be measured in terms of gallons per second, but there is no special name given to *this* unit. The international definition of the ampere is a complicated one, but the one which follows is quite correct and much easier to understand at this stage:

DEFINITION   **The unit of current called the ampere is the current that flows when an electric charge of one coulomb passes a given point in one second.**

If the charge Q coulombs flowing in t seconds is measured, then the current I amperes is given by the equation

$$I = \frac{Q}{t}.$$ Eq. 4.1

Hence if a current I flows for a time t the total charge Q which flows is given by

$$Q = It.$$ Eq. 4.2

Although the ampere (A) is of a more practical size than the **farad** (for capacitance, section 3.4.1), it is often too big for some applications, for example, for measuring the current through an X-ray tube. Then we may speak of sub-multiples of the unit; the common ones in this case are the **milliampere** (mA) and the **microampere** (μA). These units are often abbreviated in speech to 'amp', 'milliamp' and 'microamp'.

4.2   RESISTANCE AND OHM'S LAW

4.2.1   Resistance. Whenever matter moves from a higher to a lower level, it experiences **resistance** to its movement. For example, the movement of water in a stream is resisted by the irregularity of the stones over which it flows; even the falling mass in Fig. 1.1 (1.3.4) is slowed down slightly by *air* resistance. We might suspect even at this early stage that the resistance causes *loss of energy;* this is a most important point in the electrical case (5.1.1).

Similarly, when electrons flow through a conductor they experience **resistance** to their flow. This may be imagined roughly as resulting from the presence of atoms along the paths of the free electrons and the resulting collisions which take place. We shall discuss the factors influencing the resistance of a conductor in section 4.2.3.

4.2.2   Ohm's law is a very important law for electrical circuits. Fig. 4.2 shows a simple circuit which may be used to demonstrate the law. The

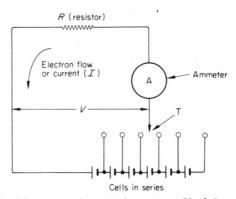

Fig. 4.2   An experiment to demonstrate Ohm's Law.

potential difference is provided by one or more cells connected *in series* (instead of only the one cell of Fig. 4.1b). These may be used to produce different p.d.s by varying the connexion T from one terminal to another. Putting cells in series (4.1.2) is equivalent to *adding* their p.d.s, just as putting distances one above the other results in their *addition.*

The circuit in Fig. 4.2 differs from Fig. 4.1b (4.1.2) in another respect. All conductors offer resistance to current flow, but to simplify circuit drawing it is usual to assume that all the resistance is concentrated in certain components called **resistors**; these are represented by zigzag lines. The lines joining the different components then represent conductors (connexions) which are imagined to have zero resistance.

In performing the experiment to demonstrate Ohm's law we vary the 'tapping point' T, thus varying the p.d. (V), and for each value we read the corresponding current (I) on the ammeter. We express the results in a table (Table 4.1). It is apparent that, within the limits of experimental error, the current (I) is proportional to the p.d. (V). This is clearly shown if we divide V by I (in the third line of the table) when we obtain a ratio which is nearly constant and whose average in this example is 4.97.

TABLE 4.1 (4.2.2) *The results of an experiment to demonstrate Ohm's law*

P.d. (V)	1	2	3	4	5	6	7	8	volts
Current (I)	0·21	0·39	0·63	0·78	0·97	1·25	1·41	1·59	amperes
$V/I (= R)$	4·8	5·1	4·8	5·1	5·2	4·8	4·95	5·03	ohms

From this experiment we can derive two pieces of information. The first is a statement of **Ohm's law**:

DEFINITION **The current through a metallic conductor is proportional to the potential difference between its ends, so long as its physical conditions remain constant.**

The second can be seen from Table 4.1; the constant value of V/I is regarded as a property of the conductor (the resistor in Fig. 4.2), and is called its **resistance** (R). The unit of resistance is called the **ohm** (Ω*).

DEFINITION **If a potential difference of one volt drives a current of one ampere through a conductor, the resistance of the conductor is said to be one ohm.**

We shall discuss resistance further in the following section, but let us first derive another equation similar to the capacitance equation (Eq. 3.1, 3.4.1). From above it is clear that $V/I = R$. Hence we can say

$$I = \frac{V}{R}, \quad \text{or} \left(V = IR, \quad \text{or} \quad R = \frac{V}{I} \right). \qquad \text{Eq. 4.3}$$

This is an algebraic expression of Ohm's law, but must not be quoted if Ohm's law is asked for in an examination. It enables us to make calculations

*Ω is the Greek capital letter omega.

of current, p.d. and resistance. For example, if V = 10 volts and R = 2 ohms, the current will be

$$I = V/R = 10/2 = 5 \text{ amperes.}$$

Again, if a current of 0·5 A is driven through a resistance of 1 000 Ω, the p.d. across the resistance will be

$$V = IR = 0.5 \times 1\ 000 = 500 \text{ V.}$$

4.2.3   Factors affecting resistance. The unit of resistance is the ohm (Ω), but, as in the case of other units, multiples are often convenient. In X-ray applications the **kilohm** (kΩ) (1 kΩ = 10^3 Ω) and the **megohm** (MΩ) (1 MΩ = 10^6 Ω) are the most useful.

What factors govern the resistance of a conductor? If we think of the resistance of a pipe through which water is flowing, we can almost guess the answer. The resistance will obviously be greater for a longer pipe, or for a thinner pipe, or for one with a rougher inside bore. If l (m) is the length of a conductor, a (m²) its cross-sectional area, and ρ* a measure of the electrical 'roughness' of its material, called its **specific resistance**, then the resistance of the conductor (R ohms) will be given by

$$R = \frac{\rho l}{a}. \qquad\qquad \text{Eq. 4.4}$$

DEFINITION   **The specific resistance of a conducting material is the resistance in ohms, measured between opposite faces, of a cube with a side of length one metre.**

Examples of the calculation of conductor resistances may be found in reference books. However, it is clear that the differences in resistance among conductors (wires) of the same dimensions but different materials result from their different specific resistances. For example, silver and copper have *low* values of ρ, whereas iron, tungsten and special alloys have relatively *high* values of ρ. These special alloys are used to make wires **(resistance wires)** which are incorporated into **resistors** when high values of resistance are required.

Another property of the specific resistance is that it is affected by temperature. In most metals, *increase* of temperature produces *increase* of specific resistance, and hence of the resistance of a wire made of that metal.

*ρ is the Greek small letter rho.

4.3   CIRCUIT LAWS

4.3.1   Combinations of p.d. and resistance. The simple circuit of Fig. 4.2 (4.2.2) illustrates that p.d.s. connected in series add together (Fig. 4.3a), so long as + is connected to − throughout the chain of cells. This is called

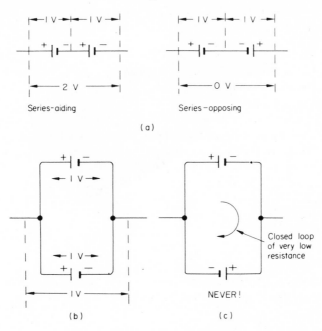

Fig. 4.3   Cells in series and in parallel.

series-aiding. Cells thus connected are often called a **battery.** Fig. 4.3a also shows what happens when cells are connected **series-opposing;** the resulting p.d. is the *difference* between the two p.d.s. and hence zero if they are equal. Fig. 4.3b shows similar cells connected **in parallel,** i.e. so that their separate *currents* add at the points of connexion. Then their combined p.d. is equal to the p.d. of each one, but twice the current can be drawn from the combination. Note that cells must *never* be connected in parallel with + to − as in Fig. 4.3c, because they are then series-aiding in a closed 'loop' of very low resistance (this is often called a **short-circuit**). As $I = V/R$, the current which would flow in the closed loop would be very large and would probably damage the cells. For the same reason, cells must never have their terminals connected together, even momentarily, by a wire of low resistance. This might be quite entertaining because it might produce a shower of sparks, but it is almost certain to damage the cell.

Figure 4.4a shows a battery connected to three resistors in parallel. In parallel means that the total current, I, divides into I_1, I_2 and I_3 through the resistors R_1, R_2 and R_3; these three currents then recombine to form the

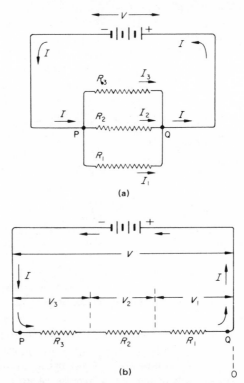

Fig. 4.4   Resistances in parallel and in series.

total current I. It is an instructive application of Ohm's law to calculate the effective resistance between points P and Q. This is done as follows. If R is the total effective resistance of R_1, R_2 and R_3 in parallel, then

$$I = I_1 + I_2 + I_3,$$

44

then as $I = V/R$,

$$\frac{V}{R} = \frac{V}{R_1} + \frac{V}{R_2} + \frac{V}{R_3},$$

then dividing by V

$$\frac{1}{R} = \frac{1}{R_1} + \frac{1}{R_2} + \frac{1}{R_3}. \qquad \text{Eq. 4.5}$$

For example, the combined resistance of resistors of 2, 3 and 4 ohms in parallel is $\frac{12}{13}$ ohm. The combined resistance is always less than that of the smallest single resistor.

Figure 4.4b shows a battery connected to three resistors in series. Here the current through all three resistors is the same, but the p.d.s add, as shown (compare cells in series). A calculation similar to the above gives us the total effective resistance between P and Q:

$$V = V_1 + V_2 + V_3,$$

then as $V = IR$

$$IR = IR_1 + IR_2 + IR_3,$$

then dividing by I

$$R = R_1 + R_2 + R_3. \qquad \text{Eq. 4.6}$$

For example, the combined resistance of resistors of 2, 3 and 4 ohms is $2 + 3 + 4 = 9$ ohms.

The circuit in Fig. 4.4b with resistors in series forms the basis of an important circuit arrangement called a **potential divider** or sometimes a **potentiometer**. It is clear that because the current through all the resistors is the same, the potential changes in proportion to the resistance value and thus increases as we progress along the resistance. Fig. 4.5a shows a practical variable potential divider, consisting of a coil of resistance wire, insulated with enamel except for a bared strip along which a contact C slides. Adjustment of C enables the p.d. V' (between C and O) to be adjusted to any desired fraction of the total, V.

Another way in which variable resistance can be of value is shown in Fig. 4.5b. Here the resistor with the sliding contact is similar to that in Fig. 4.5a, but it is connected in series with the circuit merely to act as a variable resistance (often called a **rheostat**). In this case it controls the current through a lamp to make it dimmer or brighter.

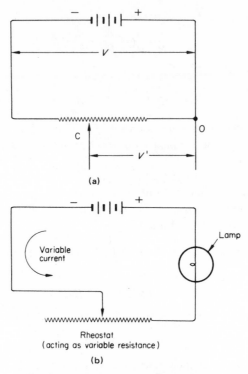

Fig. 4.5   Some uses of variable resistances.

4.3.2   Ammeters and voltmeters. Figs 4.1 (4.1.2) and 4.2 (4.2.2) show circuits which incorporate an instrument for measuring current flow, called an **ammeter**. This word is an abbreviation of ampere-meter, i.e. a meter for measuring amperes (4.1.4). A more sensitive instrument, for example one which will measure thousandths of an ampere, is called a **milliammeter**. We shall see in Chapter 6 how these instruments work, but in the meantime it is necessary to know how they are used in a circuit.

The use of an ammeter to measure the current in a circuit is simple. It is connected *in series* with the circuit so that the whole of the current to be measured flows through it (Fig. 4.2, 4.2.2). Sometimes, however, it may be necessary to alter the **sensitivity** of an instrument. For example, one may need to use a milliammeter to measure amperes; its sensitivity must therefore be effectively reduced. This is done as shown in Fig. 4.6a, by connecting in

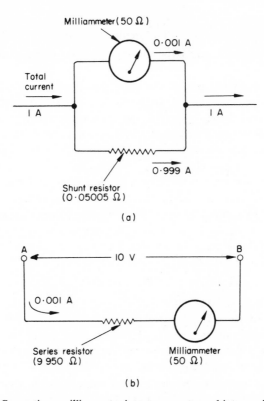

Fig. 4.6   Converting a milliammeter into an ammeter and into a voltmeter.

parallel with it a **shunt resistor,** or **shunt,** which diverts or 'shunts' a fixed fraction of the current from the meter. The resistance of the shunt is calculated as follows.

Suppose the milliammeter reads 0 to 1 mA **(full scale deflexion, f.s.d.*),** and has itself a resistance of 50 Ω. Then if it is required to read 0 to 1 A, equivalent to a sensitivity reduction of one thousand times, 1 mA (0·001 A) must flow through the meter at f.s.d., and therefore 0·999 A through the shunt. Now the p.d. across the meter, and therefore across the shunt, will be

$$V = IR = 0{\cdot}001 \times 50 = 0{\cdot}05 \text{ V (50 mV)}.$$

*The abbreviation f.s.d. is also used in radiology to mean **focus-skin-distance** but as the two applications are so different, confusion never arises.

47

Hence the resistance of the shunt will be

$$R = V/I = 0{\cdot}05/0{\cdot}999 = 0{\cdot}05005 \ \Omega.$$

In practice, the error involved in making the shunt of resistance $0{\cdot}\,05 \ \Omega$ will be negligible.

Although the resistance of the shunt has·been calculated for the f.s.d. current, this calculation will apply for all other currents. This happens because the total current always divides between the meter and the shunt in the inverse ratio of their resistances. Therefore the current through the meter, *in this example,* will always be one thousandth of the total current.

A milliammeter may also be adapted to measure the difference in potential or p.d. between two points, for example points A and B in Fig. 4.6b. The p.d. is measured in volts and is therefore often called **voltage**. The voltage is measured by measuring the current which it produces through a resistor of known value. If the milliammeter of the example above, of f.s.d. 1 mA and resistance 50 Ω, is required to give f.s.d. for a p.d. of 10 V, the total resistance must be such that 10 V drives a current of 1 mA through it. Hence

$$R = V/I = 10/0{\cdot}001 = 10\,000 \ \Omega \text{ or } 10 \ \text{k}\Omega.$$

But we have already 50 Ω in the meter. Hence the added **series resistor** should have a value of $10\,000 - 50 = 9\,950 \ \Omega$. As before, little error would result in practice from adding 10 000 Ω, which is a more readily available value of resistor than $9\,950 \ \Omega$. The meter and series resistor together form a **voltmeter**, just as the meter and shunt resistor in Fig. 4.6a together form an **ammeter**. The added resistor in both instruments is usually enclosed in the same case as the meter itself, and the scale of the meter is marked 0 to 1 A or 0 to 10 V, instead of 0 to 1 mA.

The conversion of a milliammeter to an ammeter or a voltmeter illustrates two important points. First, in performing calculations about Ohm's law it is necessary to convert everything to the standard units: amperes, volts and ohms. Thus 1 mA must be expressed as $0{\cdot}001$ A. Short-cuts to this rule do exist; for example one may work in mA, V and kΩ, but in the absence of considerable experience, mistakes may be made, and it is safer to use the standard method.

Second, it is important that a measuring instrument of any kind should not significantly disturb or alter the quantity being measured. For example, a speedometer on a car would be of little use if it absorbed a large fraction of the horse-power of the car engine! If an ammeter were not included in a circuit, it would be replaced by a conductor of practically zero resistance.

Hence an ammeter should have a very low resistance. Similarly, the points between which a voltmeter is connected normally have a high resistance between them (including any connexions which may be part of the circuit itself). Hence a voltmeter should have a very high resistance. If these conditions are not fulfilled, the use of the instruments will produce undesirable changes in the quantities they are measuring, and therefore will give erroneous values.

4.3.3. Charging and discharging a capacitor.

If the two plates of a capacitor are connected to the two terminals of a battery, negative charge flows almost instantaneously from the negative terminal of the battery into the capacitor. An equal negative charge then flows *out* of the opposite plate of the capacitor (leaving it charged equally positively) and back to the positive terminal of the battery. The capacitor is said to be **charged** and a potential difference exists between its plates equal to the e.m.f. (5.3.2) of the battery.

If now the capacitor is disconnected from the battery, it is acting as a temporary store or reservoir of electric charge; if its plates are then connected together by a conductor, the excess negative charge on one plate flows almost instantaneously round to the other plate and neutralizes the positive charge which was on it. The capacitor is then said to be **discharged** and the potential difference between its plates has fallen to zero. This storage property of capacitors is of great value in electrical circuits (for example, section 12.5.2). It can be of particular value if the charging and discharging processes are made to occur slowly and controllably; this may be done as follows.

Figure 4.7a shows a capacitor of capacitance C connected via a resistor of resistance R to a switch Sw. The switch enables the capacitor first to be connected to a battery of e.m.f. V_B, when it is *charged* gradually through the resistor, then to have its plates connected together, when it is *discharged* gradually through the resistor. Figure 4.7b shows the way in which the p.d. v_C across the capacitor varies with time. Starting with the capacitor discharged ($v_C = 0$), putting the switch in the charge position causes v_C to rise, first rapidly, then more and more slowly, until after a very long time it reaches equality with V_B. Putting the switch in the discharge position causes the reverse change in v_C until after a long time $v_C = 0$. Both these graphs represent exponential laws, the equation for the discharge being

$$v_C = V_B e^{-t/RC}, \qquad \text{Eq. 4.7}$$

where e is a mathematical constant.

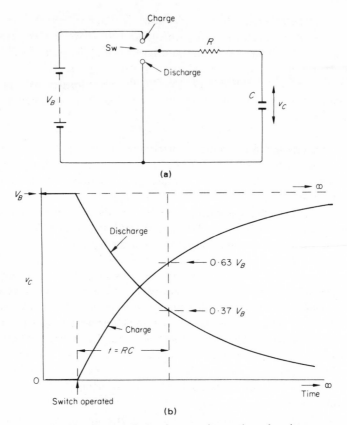

Fig. 4.7   Charging and discharging capacitance through resistance.

This equation will be recognized as similar to those governing the attenuation of homogeneous radiation in matter (14.2.3) and the decay of radioactive substances (16.5.1). The rate at which the change in v_C takes place is governed solely by the product of the resistance R and the capacitance C, viz. RC; this is known as the **time constant** of the combination. The figure shows (and it can be calculated from Eq. 4.7) that the time constant RC is equal to the time taken for v_C either to *rise* to a value of 0.63 V_B or to *fall* to 0.37 V_B.

This property of a resistance-capacitance combination enables it to be used as a kind of electrical 'hour-glass'; for example, an 'RC' circuit is used as the timing element in an electronic exposure timer (13.4.1).

50

4.3.4 Combinations of capacitances. One important example of the behaviour of a capacitor in an electrical circuit is given in section 4.3.3. Another example is as follows. In circuits, it is often necessary to connect two or more capacitors in series or in parallel; we must then be able to calculate the total effective capacitance of the combination. Similar calculations are shown in section 4.3.1 for resistors; the results for capacitors may be deduced by similar methods, but only the results will be given here.

If a number of capacitors (of capacitance C_1, C_2, C_3, etc.) are connected *in series,* the total effective capacitance C is given by:

$$\frac{1}{C} = \frac{1}{C_1} + \frac{1}{C_2} + \frac{1}{C_3} \qquad\qquad \text{Eq. 4.8}$$

(compare Eq. 4.5).

If a number of capacitors (of capacitance C_1, C_2, C_3 etc.) are connected *in parallel,* the total effective capacitance C is given by:

$$C = C_1 + C_2 + C_3 \qquad\qquad \text{Eq. 4.9}$$

(compare Eq. 4.6).

Note that the equation for capacitors *in series* is the same as that for resistors *in parallel* and *vice versa;* this arises because capacitance is a measure of the extent to which a capacitor will, in a sense, *accept* electric charge, whereas resistance is a measure of the extent to which a resistor will resist or *reject* charge.

5 Electrical energy and power

5.1 ENERGY AND POWER

5.1.1 Electrical energy. The two most important electrical quantities so far discussed have been **charge** or **quantity of electricity** (Q) (3.3.1) and **potential** (V) (3.3.3). Potential was defined in terms of **unit charge**; thus the potential at a point is equal to the work done in joules in bringing unit charge, i.e. one coulomb, from infinity to the point; in so doing we store in it a quantity of energy equal to the work done. The latter is *mechanical* (assuming that the charge has been 'pushed'), but the stored energy is *electrical,* because it arises from the repulsive force between two electric charges. We have, in fact, converted mechanical energy into **electrical energy.**

These ideas are only theoretical, and the corresponding experiment would be very difficult to do. However, the ideas lead to very important practical consequences, as follows. Suppose instead of unit charge (one coulomb) we have Q coulombs; then to bring this charge from infinity (zero potential) to a given point (potential V) would require Q times the work needed for unit charge, or VQ joules. This would result in VQ joules of electrical energy being produced. Now the potential at infinity is zero, and at the given point is V, hence the **potential difference** (p.d.) is $V - 0 = V$ volts. Therefore we can say that when a charge of Q coulombs moves through a potential difference of V volts, an amount of electrical energy given by

$$\mathcal{E} = VQ \text{ joules} \qquad \text{Eq. 5.1}$$

is involved.

There are two important practical examples of the above ideas. The first concerns capacitors (condensers). We have seen how a capacitor may be used to store electric charge for short periods, and to give it up again when required. This is called charging and discharging (4.3.3). The increase of *charge* (Q) in the capacitor, however, results also in an increase of *p.d.* (V) between the plates. The charging process is thus equivalent in a sense to bringing a charge mechanically from infinity, and results in the storage of *electrical energy* as well as *charge* in the capacitor. Conversely, when the capacitor is *discharged* (4.3.3), this energy becomes available for other purposes, either as *electrical* energy in another form or as a different *kind* of energy. For example, the spark produced when a charged capacitor is short-circuited (4.3.1) results from the *electrical* energy being suddenly converted into a spark, i.e. into *heat, light* and *sound* energy.

The second important example relates to the flow of current in a conductor, i.e. the movement of charge in a conductor. In Fig. 4.2 (4.2.2), for example, electric charge (Q) is flowing through a p.d. (V), the process involving an amount of electrical energy VQ joules. But in Fig. 4.2 we expressed the flow of charge in terms of *current* (I); how then can we calculate the energy? We can do so by recalling (4.1.4) that the ampere (the unit of current) corresponds to the number of coulombs flowing per second, viz:

$$I = \frac{Q}{t},\qquad\text{Eq. 5.2}$$

where t is the time period (in seconds) over which the current flow is considered. Hence

$$Q = It \qquad\qquad\text{Eq. 5.3}$$

and the energy $\qquad\quad\mathcal{E} = VQ \qquad$ (Eq. 5.1), or

$$\mathcal{E} = VIt \text{ joules.} \qquad\qquad\text{Eq. 5.4}$$

In words, we can say that a current I amperes flowing for a time t seconds through a p.d. of V volts involves an amount of electrical energy $\mathcal{E} = VIt$ joules.

5.1.2 Electrical power.

Let us take Eq. 5.3, $Q = It$, and divide throughout by t (time); we arrive at Eq. 5.2, $I = Q/t$. In other words, charge per second is called *current*. Now let us take Eq. 5.4, $\mathcal{E} = VIt$, and do likewise, we arrive at a new equation:

$$\frac{\mathcal{E}}{t} = VI. \qquad \text{Eq. 5.5}$$

In words, energy per second is equal to p.d. multiplied by current. The concept of energy per unit time is so important (for all forms of energy) that we give it the special name **power** (*P*). Thus in the electrical case,

$$P = VI, \qquad \text{Eq. 5.6}$$

DEFINITION   **Power is the rate at which energy is expended or converted.**

Because energy and work are equivalent, we can also say that power is the **rate of working.** In this sense the word power comes very close to one of its everyday meanings. For example, power was originally expressed in terms of the rate at which an average horse of a given breed and size could comfortably work; this rate is called **one horse-power.** As we know, the rate at which motor-car engines will work, or will produce mechanical energy, which is the same thing, is still expressed in horse-power. For scientific purposes however the unit is differently defined but is obsolescent.

The S.I. unit of energy is the **joule** (1.3.4), therefore **power** *could* be expressed as joules per second. But power is such an important concept that, as in the analogous case of electric current (4.1.4), its unit is given a special name: the **watt (W).**

DEFINITION   **The unit of power called the watt is equal to one joule per second.**

If the energy flowing in *t* seconds is measured, then the power *P* is given by the equation

$$P = \frac{\mathcal{E}}{t}. \qquad \text{Eq. 5.7}$$

Hence if a power *P* continues for a time *t* the total energy is given by

$$\mathcal{E} = Pt. \qquad \text{Eq. 5.8}$$

It is also important in electrical circuits to realize clearly the implications of Eq. 5.6, viz. that **power (watts)** is equal to the product of **p.d. (volts)** and **current (amperes).**

As already explained, the S.I. unit of energy is the joule. However, for commercial purposes, electrical energy is sold in terms of a different unit, called the Board of Trade Unit (B. of T. Unit). This is the 'unit' which appears on our electricity bill every quarter; it is identical with **one kilowatt-hour** (1 kWh) and is quite simply related to the joule as follows.

The B. of T. Unit = 1 kilowatt for 1 hour
                  = 1 000 watts for 3 600 seconds
                  = 3 600 000 watt seconds.

Therefore 1 B. of T. Unit = 3 600 000 joules.

5.1.3 Numerical examples illustrating p.d., charge, current, resistance, energy and power.

Example A. An electric lamp (which can be regarded as equivalent to a resistor) has a resistance of 200 Ω, and is designed to operate with a p.d. of 100 V across its terminals. Calculate

  (i)   the current flowing through it,
  (ii) the charge flowing through it in a period of 10 s,
  (iii) the energy it consumes in 20 s, and
  (iv) the power consumption.

(i) From Ohm's law (4.2.2, Eq. 4.3), $I = V/R$; therefore $I = 100/200 = 0.5$ A.

(ii) One ampere is equivalent to one coulomb per second (4.1.4), hence from Eq. 5.3 (5.1.1): $Q = It = 0.5 \times 10 = 5$ coulombs in 10 seconds.

(iii) In 20 seconds the charge $Q = It = 0.5 \times 20 = 10$ coulombs. Now one coulomb flowing through a p.d. of one volt corresponds to an energy of one joule (5.1.1), hence from Eq. 5.1 (5.1.1): $\mathcal{E} = VQ = 100 \times 10 = 1\ 000$ joules. (If we wished, we could call this one *kilo*-joule, although this word is not often used.)

(iv) There are two methods of calculating the power consumption of the lamp.

The first method follows from (iii) and is an indirect method: power is equivalent to energy per second (5.1.2), hence from Eq. 5.7 (5.1.2): $P = \mathcal{E}/t = 1\ 000/20 = 50$ watts.

The second method is more direct and is nearly always used (with variations to be explained in section 5.2.1) when calculating power in an electrical circuit. Eq. 5.6 (5.1.2) shows that power is equal to the product of p.d. and current. Hence $P = VI = 100 \times 0.5 = 50$ watts. We note with satisfaction that both methods give the same answer!

Example B. An electric heater (which can likewise be regarded as a resistor) when operated normally with a p.d. of 250 V consumes a power of 3 kW (kilowatts). Calculate

  (i) the current flowing through it,
 (ii) its resistance,
(iii) the power it would consume if the p.d. fell to 200 V (assuming the resistance to remain constant), and
 (iv) how many B. of T. Units it would consume in 10 hours.

  (i) From Eq. 5.6 (5.1.2), $P = VI$. Hence $I = P/V = 3\ 000/250 = 12$ A. (Notice that it is essential to convert multiples or submultiples such as kW, mA, etc., to the basic units, or great errors may result.)
 (ii) From Eq. 4.3 (4.2.2), $R = V/I = 250/12 = 20\frac{5}{6}$ Ω.
(iii) If the resistance remains constant ($20\frac{5}{6}$ Ω) the result of a reduced p.d. is to cause the current to become less. This can be calculated from Eq. 4.3: $I = V/R = 200/20\frac{5}{6} = 9\frac{3}{5}$ A.

Hence the new power (Eq. 5.6) is $P = 200 \times 9\frac{3}{5} = 1\ 920$ watts or $1·92$ kW. A simpler method of performing such a calculation will be shown in section 5.2.1 (Eq. 5.8).

 (iv) A B. of T. Unit is equivalent to one kilowatt-hour. Therefore 3 kilowatts for 10 hours is 30 kilowatt-hours or 30 'units'.

5.1.4 A simple charge, current, energy and power diagram. Fig. 5.1 may assist the reader to visualize and to memorize the relationships between the above quantities.

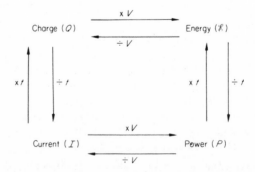

Fig. 5.1   A simple charge, current, energy and power diagram.

5.1.5 The 'effects' of an electric current; conversions of energy. We have seen (4.1.1) how a flow of electrons along a wire (conductor) constitutes an **electric current**. When such a current flows, certain 'effects' may be observed if the conditions are favourable. For example, one of the best known and important effects for our purpose is the production of heat in the conductor. This is often called the **heating effect of a current**, and it can be imagined to take place as follows. The electrons constituting the electric current flow because they receive energy (electrical) from the difference of potential (p.d.) between the ends of the conductor. In flowing between the atoms or molecules of the conductor, the electrons progressively impart this energy (electrical) to the atoms or molecules, resulting in an increase in their mechanical vibrational energy. This corresponds to a rise in **temperature** of the conductor (1.4.1) and therefore an increase in **heat** energy (1.4.2). Thus the 'heating effect of a current' really consists of a conversion of *electrical* energy to *heat* energy; it will be discussed in greater detail in section 5.2.1.

Similarly, if an electric current is passed through a conducting solution, for example water to which a little acid has been added, the flow of current will cause the water to be separated into its two constituents, hydrogen and oxygen. This is one example of the **chemical effect of a current**, and, like the heating effect, is fundamentally an energy conversion. Here, the *electrical* energy is converted into *chemical* energy. This can be shown in a spectacular fashion by mixing the resulting hydrogen and oxygen and setting light to the mixture. A loud explosion and flash result (the chemical energy is converted to heat, light and sound energy) and water is reformed. The chemical effect is not used in radiology proper, but affords one valuable method of recovering silver from dark-room fixing solutions.

The third important effect of an electric current is the **magnetic effect**. The simple phenomena of magnetism are familiar in everyday life; for example, it is well known that a magnetic compass points to the north because it is aligned into a north-south position by the magnetic forces of attraction of the earth. Nevertheless, the magnetic effect of a current is not so easy to explain in terms of energy conversions because it involves the somewhat mysterious phenomenon of magnetism. The latter, together with its relation to electric currents, will be explained in Chapter 6.

5.2 THE HEATING EFFECT OF A CURRENT

5.2.1 The conversion of electrical energy into heat; efficiency. When a quantity of electricity (charge) flows through a conductor having resistance, a certain amount of electrical energy is involved in the flow (5.1.1). In section 5.1.5 we explained that this *electrical* energy is in fact converted into *heat* energy; it will raise the temperature of the conductor if the conditions are appropriate. The amount of electrical energy is given by Eq. 5.1 (5.1.1): $\mathscr{E} = VQ$ joules. Because the joule is also the S.I. unit of heat, it would be reasonable to say that the amount of *heat* energy *also* is given by $\mathscr{E} = VQ$ joules. This statement is true only when *all* the electrical energy is converted into heat.

So far, we have described energy conversion only in *qualitative* terms. However, it is very important to discuss its *quantitative* implications (1.2.1), because energy costs money and effort. When one form of energy (A) is converted into another form (B), all of A may not change into B. If it does, then we say that the conversion process has an **efficiency** or 1·00 (unity) (or 100%, whichever is preferred). If, however, less of B is produced than the amount of A which disappeared, the process has an efficiency of less than 1·00.

DEFINITION   **The efficiency of an energy conversion process is equal to the amount got out divided by the amount put in.**

In view of the law of **conservation of energy** (1.1.4), the reader may wonder how any energy conversion process can possibly have an efficiency of less than 1·00. However, we are considering the conversion of A into B only; if a smaller amount of B appears than of A disappears, some other forms of energy C, D, etc. must have been produced, so that A = B + C + D, etc. For example, the electrical energy supplied to a radiant electric fire (5.2.2) is converted mostly to heat but also partly to light.

It so happens that most forms of energy change very easily into heat; in particular, the efficiency of conversion of electrical energy into heat is very nearly 1·00. Hence we can make the statement above that the amount of heat energy is given by $\mathscr{E} = VQ$ joules.

This equation is not in a very useful form, because normally we know the *current* in a circuit and not the charge. But from Eq. 5.4 (5.1.1), $\mathscr{E} = VIt$ joules, hence if a p.d. V volts drives a current I amperes through a resistance R ohms for t seconds, the amount of heat produced will be VIt joules. This equation is useful only if V and I are known; however, from Ohm's law (Eq. 4.3, 4.2.2) we can deduce also that

$$\mathscr{E} = \frac{V^2 t}{R} \text{ joules} \qquad\qquad \text{Eq. 5.9}$$

or

$$\mathcal{E} = I^2Rt \text{ joules.} \qquad\qquad \text{Eq. 5.10}$$

By using one of these equations or the original Eq. 5.4 (5.1.1), $\mathcal{E} = VIt$ joules, we can calculate the heat produced in a given time in any electrical circuit, *in joules*.

When discussing heat and its units (1.4.2) we pointed out that the **calorie** (cal) was still widely used as a unit of heat, also that 1 calorie = about 4·2 joules. Hence we can say that the amount of heat, H, is given by

$$H = \frac{VIt}{4\cdot2} \text{ cal,} \qquad\qquad \text{Eq. 5.11}$$

or

$$H = \frac{V^2t}{4\cdot2R} \text{ cal,} \qquad\qquad \text{Eq. 5.12}$$

or

$$H = \frac{I^2Rt}{4\cdot2} \text{ cal.} \qquad\qquad \text{Eq. 5.13}$$

These equations enable us to calculate the heat produced in a given time in any electrical device, for example in an X-ray tube. Given its thermal capacity (1.4.2) we can then calculate its temperature rise (assuming that it loses no heat by conduction, etc.).

5.2.2 Some applications of the conversion of electrical energy to heat. The ease and efficiency with which electrical energy can be converted into heat can be either an advantage or a disadvantage.

The most important disadvantage in radiology is to be found in the X-ray tube itself. As will be explained in Chapter 10, the object of the X-ray tube is primarily to convert electrical energy (A) into X-ray energy (B). However, the efficiency of this process is exceedingly low, usually less than 1% (10.3), and most of A is converted into heat, i.e. C (5.2.1) with an efficiency of *nearly* 100%. This has the double disadvantage that for a given amount of electrical energy, not only is very little X-ray energy produced, but also the very large amount of heat which is a by-product is very difficult to get rid of. A numerical example will show how large is this effect; it will also illustrate the theory in section 5.2.1.

Suppose an X-ray tube operates with a p.d. of 250 kV and a current of 15 mA. It is required to calculate how much heat is produced in one minute of operation. (This example is typical of a 'deep' X-ray therapy treatment

59

exposure.) From Eq. 5.6 (5.1.2), the electrical power $P = VI =$ 250 000 x 0·015 = 3 750 watts. Because nearly all the electrical energy is converted into heat, we can see immediately that the tube will produce more heat per second than a large electric fire. Now from Eq. 5.8 (5.1.2), $\& = Pt = 3\ 750 \times 60 = 225\ 000$ joules. This is the heat produced. If the result is wanted in calories, $H = 225\ 000/4·2 = 53\ 500$ cal. Of course, this result could have been obtained directly by using Eq. 5.11 (5.2.1).

The following are examples in radiology in which the heating effect of a current is advantageous. First, the effect enables heat to be produced very simply wherever it is wanted. We all appreciate the convenience of a radiant electric fire, which can be easily carried around and the heat directed where required. The same principle is adopted in X-ray tubes and valves. In these, for reasons explained in Chapter 11, it is necessary to heat a conductor to white heat. This is done by passing a current through a very thin conductor, usually made of **tungsten**, called the **filament**. This is enclosed in a glass bulb from which all the air has been removed. The heat produced by the current cannot easily escape from the filament (only by **radiation**, 1.4.3), and the tungsten filament glows white hot. Imagine how awkward it would be if we had to produce the heat by means of a gas flame!

The same principle operates in the ordinary filament electric lamp (not the fluorescent kind). In this case, however, the glass bulb is filled with an inactive gas (usually argon) at low pressure; this enables the filament to be heated to a higher temperature without damage and to give off a lot of brilliant white rather than red light.

Another example of the electrical production of heat is very different from the above. In all electrical circuits, for example in a house or in an X-ray generator, there is a danger that such a large current may flow (because of some fault) that the conductors become overheated and perhaps melt or cause a fire. This danger is avoided by including in series with the circuit a short link of much thinner wire, called a **fuse**, suitably supported in a fireproof holder. Besides being thinner, the wire might also be made of a material, such as tin, with a relatively low melting point. Then if the current in the circuit rises to a dangerous level, the fuse melts ('blows') and 'breaks' the circuit, thus preventing further current flow. Of course, the apparatus then ceases to function, but this is better than having a fire, and the fault can be found and corrected. If a fuse 'blows', another must never be substituted before finding out why the first one 'blew'. Examples of fuses will be found in the circuits of Chapter 13.

5.3 SOURCES OF ELECTRICAL ENERGY; E.M.F. AND P.D.

5.3.1 Sources of electrical energy. Whenever one wishes to have power or energy of a particular form, one must usually obtain it by conversion from another form. Thus there are many varieties of sources of electrical energy; each has its different sphere of application.

Perhaps the simplest way of obtaining electrical energy is to convert it from chemical energy. We have already seen a very elementary example of this in the **simple cell** of sections 4.1.2 and 4.1.3, and Fig. 4.1. There it was explained that the continued conversion of chemical into electrical energy results in the cell possessing an **electromotive force** (e.m.f.). This e.m.f. enables the cell to maintain a **potential difference** (p.d.) across its terminals; the p.d. maintains a flow of current through a conductor connecting them. The precise difference and relationship between e.m.f. and p.d. will be explained in section 5.3.2; meanwhile we shall describe some other more practical sources of electrical energy.

The simple cell is the prototype of a large range of devices in which chemical energy is converted into electrical energy. These devices can be divided broadly into: (a) **dry batteries** and (b) **accumulators** or **storage batteries**. In dry batteries, when the chemical energy in the battery is exhausted, a new battery must be substituted. This type is familiar in pocket torches, transistor radios, etc. In accumulators or storage batteries, when the chemical energy is exhausted the device is said to be **discharged**; it can be replenished by reversing the original chemical changes. This is done by passing a current in reverse through the accumulator, a process known as **charging**. The common 'car battery' is an example of an accumulator. The charging and discharging processes are reminiscent of those encountered in the *temporary* storage of electric charge in a capacitor (3.4.1). In the latter, however, it is the *charge itself* which is stored, whereas in accumulators only *energy* is stored in *chemical* form and reconverted when required. The accumulator or storage battery offers the best method of 'storing electrical energy' for *long* periods. Dry batteries and accumulators are little used in radiology, mainly because chemical energy is *very* expensive and much cheaper and more convenient sources of electrical energy are available.

The best and cheapest way of producing large amounts of electrical energy is to convert it from *mechanical* energy. This is done in all modern power stations, and thence the electrical energy is fed through cables as **alternating current** to hospitals, factories, houses, etc. (8.1.2). The mechanical energy is

61

obtained in a variety of ways, from a steam engine powered by coal or nuclear fuel, from a diesel engine or from water power in countries where it is plentiful. To perform the conversion, a *link* between the mechanical effect and the electrical effect must be found. This link is **magnetism**, discussed in Chapter 6. The mechanical energy is fed into machines called **alternators** (8.1.2) which by virtue of the magnetic link produce electrical energy in a particularly convenient form called **alternating current**. Indeed, as is well known, it is possible to perform the reverse operation. Electric current can be passed through a machine similar to an alternator, called an **electric motor**, and mechanical energy derived from it, e.g. in an electric train or in a vacuum cleaner (6.3.3).

5.3.2 E.m.f., p.d. and internal resistance. The simple cell of Fig. 4.1 (4.1.2) is a 'generator' or converter of electrical energy that exhibits fundamental properties common to all types of energy converter. If we study these properties for the simple cell we shall learn a great deal that will explain many of the phenomena, to be described later, which are encountered in X-ray engineering.

The ability of a cell (or any generator) to maintain a **potential difference** (p.d.) between its terminals arises from the conversion of chemical (or other) energy into electrical energy. This ability is measured by its **electromotive force** (e.m.f.) (4.1.3). Both **e.m.f.** and **p.d.** are measured in **volts**; the distinction between them is not always clear, so that very often one uses the colloquial term **voltage**. This may often conceal the possibility that one does not know which is which!

Equation 5.1 (5.1.1) tells us that energy is equal to the product of p.d. and charge ($\mathcal{E} = VQ$). Hence $V = \mathcal{E}/Q$; from this we may deduce that if a cell can produce an amount of electrical energy $\mathcal{E}$ joules while a charge Q coulombs passes through it, the charge has its *potential* raised by V volts. This rise in potential is in fact the maximum rise the cell will produce under any circumstances, and is called its *e.m.f.* In this context the cell may be compared to a water pump which converts mechanical energy from its driving force to potential energy of the water pumped out. If one imagines a kind of 'e.m.f.' of the pump, this would be equivalent to the maximum height to which the pump could ever drive the water.

In practice, the cell (or other generator) will rarely be able to produce a *p.d.* which is as large as its *e.m.f.* The p.d. across the cell terminals is usually smaller than the e.m.f., and becomes smaller the greater the current which the cell is called upon to deliver. The reader will probably be very familiar with this type of behaviour, perhaps without realizing its cause. For example, if one switches on a large electric fire (drawing a large current), the electric light in the room will be seen to become slightly dimmer. Similarly, while the starter of a motor car is pressed, a very large current is drawn from the car battery (accumulator), causing the car lights to dim considerably. Clearly, the increased current has caused a fall in p.d. Of course, it is easy to dismiss this behaviour by saying that drawing more current for the electric fire or the starter has left less for the lights, but this is not a precise explanation. To understand the phenomenon fully we must consider the composition of the cell (or other generator) in more detail.

A cell (generator) has within it a source of e.m.f. But a *practical* cell is also made up of conductors and chemical solutions, etc. It therefore possesses *resistance*; this is known as its **internal resistance** and is a very important property. Its effect on the behaviour of the cell (generator) is illustrated in Fig. 5.2. Here the cell (generator) is shown as a container (in dashed lines)

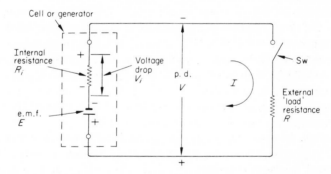

Fig. 5.2   The e.m.f., p.d. and internal resistance of a cell or generator.

with two terminals; within the container, between the terminals, is a cell symbol representing the e.m.f. (E), and a resistor symbol representing the internal resistance (R_i) in series. Of course, in the actual cell E and R_i do not exist in such convenient separate 'lumps'; they are distributed and mixed together in small parts throughout the cell. However, in a simple analysis they can be treated as though they were simple components, as shown.

Now let us suppose that the switch Sw is 'open', that is its contacts are separated and it will not allow the flow of current. Hence the current I

throughout the whole circuit, including the internal resistance R_i, is zero. From Fig. 5.2, V_i is the p.d. caused by the flow of current through R_i; it is given by $V_i = IR_i$ (Eq. 4.3, 4.2.2). As I is zero, V_i must also be zero, whatever the value of R_i. Hence the p.d. across the cell terminals, V, which is equal to $(E - V_i)$, is $(E - 0)$ or simply E. Therefore we may conclude that when a cell (generator) is delivering no current its **terminal p.d.** is a maximum and is equal to its e.m.f.

Now suppose that the switch is 'closed'; its contacts are brought together and current can flow. Let us try to calculate the terminal p.d. with a current I flowing. The current I flows through the internal resistance R_i; it therefore produces a p.d. across the resistance given by $V_i = IR_i$ (Eq. 4.3). Then, as shown above, the terminal p.d. is given by $V = E - V_i$. Hence

$$V = E - IR_i. \qquad \text{Eq. 5.14}$$

This equation expresses quantitatively the type of behaviour discussed qualitatively above. For example, let us consider a car battery of $E = 12$ V, and internal resistance $0.01\ \Omega$. If no current is drawn from it, $I = 0$, $V_i = 0$ and the terminal p.d. is equal to the e.m.f., i.e. 12 V. Now suppose we switch on the headlamps which draw a current of 10 A. Then $V = 12 - (10 \times 0.01) = 11.9$ V. The terminal p.d. has fallen by only 0.1 V, i.e. scarcely at all. Now let us press the starter button, causing an additional current of 200 A to flow (it takes a lot of power to start a car engine, as we know to our cost if we have to 'crank' it!). Then $V = 12 - (210 \times 0.01) = 9.9$ V. The terminal p.d. has dropped by a further 2 V, i.e. enough to cause the lights to dim appreciably.

The whole explanation of this type of behaviour, which is very important in X-ray engineering, lies in the p.d. produced by the flow of current through the internal resistance. This kind of p.d. is called **voltage drop**; it appears in many different ways in X-ray generators and we shall discuss it again in Chapters 12 and 13.

6 Magnetism and electricity

6.1 MAGNETISM

6.1.1 Magnetism; the laws of magnetic force. We are all familiar with the **magnetic compass**, which is used in ships, in aircraft and by ramblers to indicate the north–south direction. (It is quite different, of course, from the instrument used for drawing circles, which is more properly referred to as a 'pair of compasses'.) The magnetic compass contains a bar of metal called the **compass needle** pivoted at its centre on a sharp point, and possessing the property which we call **magnetism**.

Like electricity (3.1.1), it is difficult to *explain* the phenomenon of magnetism, but we can convey some idea of its nature by describing some of its properties. We also show (6.2) that it has important associations with electricity.

The compass needle is one example of a **magnet**. It is called a **permanent magnet** because it retains its magnetism, not necessarily for ever, but at least for a very long time. When mounted as in the magnetic compass, the magnet always points in a north–south direction; moreover, *the same end* always points north. This is called the **north-seeking pole** or **north pole** of the magnet, the other end being the **south-seeking pole** or **south pole**. Why does the compass needle behave like this?

If we take another permanent magnet in the form of a bar or rod, called a **bar magnet,** we shall naturally find that it behaves like a compass needle if suspended horizontally by a thread. We shall also find that its north-seeking pole *repels* the north pole of the compass needle and *attracts* the south pole

(Fig. 6.1). This behaviour recalls the law of *electric* force (3.1.2), except that positive and negative *electric charges* can exist separately, whereas north and south *magnetic poles* always go in pairs. The **law of magnetic force** states:

DEFINITION   **Similar magnetic poles repel, dissimilar magnetic poles attract.**

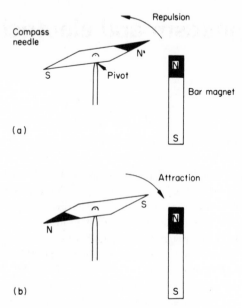

Fig. 6.1   The laws of magnetic force.

The compass needle acts as it does because the earth *behaves* as though it has a large bar magnet lying in it, very roughly coincident with its axis of rotation and with its poles near the geographical Poles of the earth. Of course, the earth's magnet does not exist in this simple form; the true explanation of the earth's magnetism is not known with certainty.

6.1.2   Magnetic induction; the molecular theory of magnetism. A magnet will attract not only other *permanent* magnets; it will also attract unmagnetized objects so long as they are made of **magnetic materials**, i.e. materials which are capable of being magnetized. Examples of such materials are some steels, iron, nickel and a number of special alloys. An instructive experiment is to dip a bar magnet into a pile of iron filings. The magnet will pick up the filings, which will adhere mostly to the *ends* of the magnet (Fig. 6.2). This shows that the *magnetism* is concentrated near the ends of the magnet; these, as explained above, are called its **poles**.

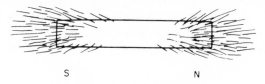

S   N

Fig. 6.2   The poles of a magnet.

The reader may wonder why, in view of the law of magnetic force (6.1.1), a magnet will attract an apparently *unmagnetized* object. A similar phenomenon exists in electricity, and was shown (3.2.2) to depend on the phenomenon of **electric induction**. There is a precisely similar effect called **magnetic induction**; this causes the permanent magnet first to magnetize the iron filings by magnetic induction, so that each filing becomes a magnet with two poles. These tiny magnets are then attracted to the large magnet.

If the iron filings in the above experiment are scraped off the magnet, then shaken about a little or heated, they will quickly lose their magnetism. They are therefore called *temporary* magnets. Some magnetic materials such as certain steels and other alloys, once magnetized, will retain their magnetism for a long period. Other materials such as soft iron and other alloys can quickly become demagnetized.

In this book we shall not be concerned very much with fundamental theories; however, a simple theory, called the **molecular theory of magnetism**, will assist us to understand many magnetic effects. In brief, the molecular theory states that magnetic materials are made up of tiny, sub-microscopic (molecular) magnets, each with a north and a south pole. These molecular magnets are more or less free to turn. If the substance is *unmagnetized*, the molecular magnets are assumed to lie *randomly* in the large magnet so that they do not reinforce each other. If the substance is *magnetized*, however, the molecular magnets are assumed to lie in *regular* formation, viz. NS–NS–NS, so that their magnetic effects all add.

This theory explains many phenomena, such as the making of magnets (explained in more comprehensive textbooks), magnetic induction, etc. It also offers an explanation of the differences between magnetic materials:

permanent magnets are those whose molecular magnets are not very free to turn, whereas temporary magnetic materials such as soft iron have molecular magnets which turn very freely. Moreover, it can be seen that shaking, hammering or heating tends to demagnetize a material by causing the molecular magnets to lie randomly. Of course, one is ultimately obliged to give an explanation for the molecular magnets; it is thought that these are associated with the atoms or molecules of the material themselves, and that the magnetism results from the movement of electric charges in the atoms (6.2).

6.1.3 Magnetic fields. As with electric and gravitational forces (3.1.2), magnetic forces can be thought of in terms of **magnetic fields**; a magnetic field is simply a region around a magnetic pole in which magnetic force is exerted on another pole, which is itself surrounded by a magnetic field. Alternatively, we can say that the two magnetic fields interact, resulting in a force of attraction (or repulsion). For our present purpose, the concept of the magnetic field is useful, so we shall discuss it in some detail.

Figure 6.3 shows a bar magnet with its N and S poles. We wish to investigate the form of the magnetic field around the magnet. A simple way of doing this is to cover the magnet with a sheet of card (through which the magnetic force passes unhindered) and to sprinkle iron filings on the card. When the card is tapped gently the iron filings take up the shape of the

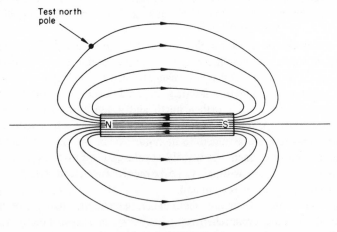

Fig. 6.3   The magnetic field of a bar magnet.

magnetic field, because the filings each turn to lie in the direction of the magnetic force. A more accurate and instructive way of 'mapping' the magnetic field is to put a freely moving north pole (assumed for this purpose to be a long way from its attendant south pole) in the field and to mark out the different paths it traces in being repelled from the N to the S pole of the bar magnet. The result of this admittedly difficult experiment is shown in Fig. 6.3. The magnetic field consists of an infinite number of possible paths of the 'test' N pole, from N to S through the air. These paths are in fact completed through the magnet itself from S to N (as shown), but in practice the test N pole would not of course be able to move through the magnet.

All the possible paths of the test N pole are called magnetic lines of force; these lines (like lines of latitude and longitude) have no physical existence, of course, but merely assist one to visualize the form of the magnetic field. Note that the magnetic field is a *vector* quantity (1.3.2); its *direction* is indicated by the direction of movement of a test *north* pole.

The shape of the magnetic field around a bar magnet is not of great interest to us except as an example. Later we shall encounter other forms of magnetic field; however, an important one is shown as a second example in Fig. 6.4. Here the magnet is like a bar magnet bent nearly into the form of a

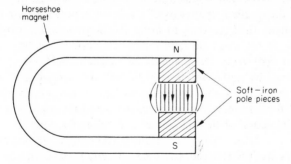

Fig. 6.4   The production of a strong, parallel magnetic field.

horseshoe, so that the N and S poles are brought fairly close together. The purpose of this arrangement is to increase the *strength* of the magnetic field between the poles; this happens because a magnetic field exists more easily in a magnetic material than in air. We say that the *permeability* of the iron is greater than that of the air (6.2.3). Hence shortening the length of the air path strengthens the field.

In this particular example we also wish to *shape* the magnetic field so that its lines of force lie exactly parallel over a considerable volume. This is done by fitting to the ends of the horseshoe suitably shaped blocks of soft iron, called **pole-pieces**. The pole-pieces in Fig. 6.4 produce *parallel* lines of force between them, as shown. We shall make use of the idea of **parallel magnetic fields** in later sections (6.2.2, 7.1.3).

6.2   THE MAGNETIC EFFECT OF AN ELECTRIC CURRENT

6.2.1   Stationary conductors; the motor effect. One aspect of the link between electricity and magnetism is demonstrated by **Oersted's experiment**. A compass needle, pointing north-south, has a conductor supported *above* it, also north-south, close to it but not touching (Fig. 6.5a). If an electric current is passed through the conductor, the compass needle will be deflected in a certain direction (Fig. 6.5b). If now the conductor is supported *below* the needle, the deflexion of the compass needle will be reversed (Fig. 6.5c). How can these effects be explained?

First, it is clear that the flow of electrons is exerting a magnetic force on the compass needle and is therefore producing a magnetic field. The final position of the needle is the result of a balance between the strength of the earth's magnetic field (tending to make the needle point north–south) and the strength of the magnetic field due to the current (tending to make the needle point east–west). The shape of the magnetic field due to the current may be guessed by noting the reversal of deflexion according to whether the needle is above or below the conductor; the field is in fact cylindrical in form, the lines of force forming concentric circles about the conductor. Consideration of the direction of the lines above and below the conductor (Fig. 6.5d) shows why the deflexion of the needle is reversed. If the direction of the electron flow itself is reversed, it can be correctly guessed that the deflexions will again reverse as a result.

In this simple apparatus we have in fact a very crude example of a current-measuring instrument, yet one which embodies all the basic features required, viz. an *indicating device* (the needle), something (the conductor) to make the current produce a *proportional force,* and a *restoring force* (from the earth's magnetic field) to balance the force due to the current. The degree of deflexion of the compass needle could be measured on a scale and could indicate the value of the current flowing in the conductor. The principles of

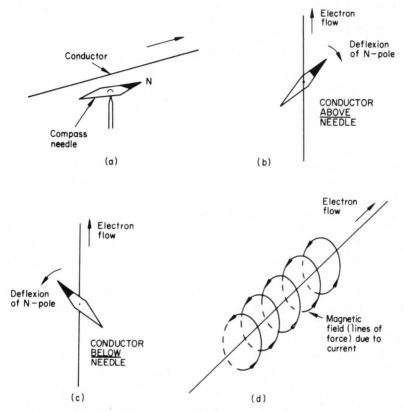

Fig. 6.5   The magnetic field due to an electric current.

more advanced measuring instruments will be explained in sections 6.3.1 and 6.3.2.

It is very instructive to examine Oersted's experiment in more detail. One might be tempted to say (correctly) that *electricity* is being converted into *magnetism*. However, it is not true to go a stage further and to claim that *electrical energy* is being continuously converted into *magnetic energy* (even if such a thing as the latter should exist). It is not clear from this experiment why this interpretation is not correct; it will be explained in section 7.1.1. At present we can only say that *electrical* energy is being converted into *mechanical* energy, the latter being associated with the *movement* of the compass needle against the *restoring force* of the earth's magnetic field. The magnetic field of the current merely acts as a *link,* enabling the energy of the current to be converted to the potential energy stored in the compass needle in its new position in the earth's magnetic field. Because this enables an electric current to produce mechanical *movement,* the effect is called the **motor effect** of a current. The remainder of this chapter is devoted to examples and important applications of the motor effect.

6.2.2 Moving conductors; Fleming's left-hand rule. If we push against a brick wall, we naturally expect that we shall push ourselves away from it. If we were standing very firmly, and the wall were very weak, then it is conceivable that the wall would move! Similarly, in Oersted's experiment, if the compass needle were fixed firmly, we should expect the conductor to move if it were very light and freely suspended. This experiment would be almost impossible to perform because the magnetic field of the compass needle is so weak. However, if we use a strong parallel magnetic field associated with a fixed magnet, as in Fig. 6.4 (6.1.3), we can demonstrate the motor effect in another form, viz. **the mechanical force on a conductor carrying current in a magnetic field.**

Figure 6.6a shows the arrangement in perspective; the magnetic field runs from left to right between the pole-pieces, and the conductor lies at right-angles to the magnetic field. The conductor can be connected via a switch to a battery so as to drive a current through it in either direction.

If the switch in Fig. 6.6a is closed, current will flow as shown. (For reasons given later *conventional current* is indicated, i.e. the opposite of *electron flow.*) The current will produce its associated magnetic field; this will interact with the existing fixed magnetic field to produce a relative force between the conductor and the magnet. In this case it is the conductor that moves, and it can be deduced from the form of the two fields that it will move at *right-angles to both the current and the magnetic field.* This is another example of the **motor effect**; it is more important than the previous example because it leads to more efficient **electromechanical** devices (6.3.2, 6.3.3). We must again emphasize that in the motor effect *electrical* energy is being converted into *mechanical* energy, the magnetic field acting as a *link* between the two. Of course, if the conductor is rigidly held then no *movement* will take place (and no *energy* will be converted), but a *force* will still be produced at right-angles to the current and to the field.

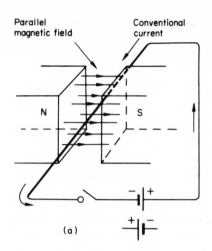

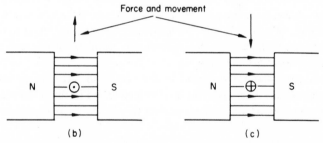

Fig. 6.6 The force on a conductor carrying a current in a magnetic field.

Whether the movement or force will be upwards or downwards can also be deduced from first principles, but there exists a simple rule, called **Fleming's left-hand rule**, by which this can easily be predicted. To apply this rule, one's left hand is arranged so that the thumb, the first finger and the second finger are all lying at right-angles to each other (i.e. along adjacent edges of a cube) (Fig. 6.7). These represent the three factors according to the following memory-aid:

> First finger denotes the magnetic Field
> seCond finger denotes the conventional Current, and
> thuMb denotes the Movement (or force).

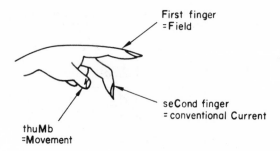

First finger
=Field

seCond finger
= conventional Current

thuMb
=Movement

Fig. 6.7   Fleming's left-hand rule.

Figures 6.6b and c show two examples of the application of this rule. The conductor is shown in much enlarged cross-section; the *dot* indicates conventional current coming *out* of the paper, and the *cross*, current going *into* the paper (standing for the *point* and the *feathers* respectively of imaginary arrows travelling in the two directions). One can easily deduce from the rule that if only one of the factors is reversed, the movement or force reverses, whereas if two factors are simultaneously reversed, the movement or force is unaltered.

It may seem unimportant to know the precise direction of movement of the conductor, but when this rule is combined with Fleming's right-hand rule (7.1.3) a very important principle results which will be explained in section 7.1.4.

6.2.3   Solenoids and permeability; relays and contactors. The magnetic field of a single conductor, even with a large current flowing, is relatively weak. The field can be made stronger by placing many conductors side-by-side, all carrying the same current. This is most conveniently done by winding insulated copper wire around a cylindrical tube, thus forming a **coil** of wire which may have many layers of wire and many thousands of turns. Such a coil is called a **solenoid**; a simple example with a single layer of wire is sketched in Fig. 6.8a. It is not difficult to see that the magnetic fields of all the separate coils will add together; in fact the field of the whole solenoid closely resembles that of a bar magnet (Fig. 6.3, 6.1.3). However, it differs from a bar magnet in that the magnetic field *appears* and *disappears* as the current is switched *on* and *off* respectively.

The magnetic field due to a current can be increased, not only by increasing the number of conductors, but also by including a magnetic material such as soft iron in the field. A common arrangement is a solenoid with a soft-iron **core** (Fig. 6.8b). The magnetic field of the solenoid magnetizes the soft iron; the magnetic field due to the iron then adds to the original field which produced it. The result is many hundred or thousand times as strong as the original field because the field due to the core is much stronger than the field which produced it. This arrangement is in fact a controllable magnet; it is called an **electromagnet.**

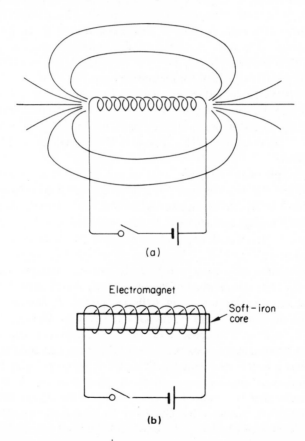

(a)

Electromagnet

Soft – iron
core

(b)

Fig. 6.8   The magnetic field of a solenoid, and an electromagnet.

The effect of the **soft-iron core** in increasing the magnetic field is a very important one in electrical apparatus. If H represents the strength of the coil's magnetic field alone, and B the strength of the field it produces in the iron core, then the ratio B/H is called the **permeability** μ of the iron (6.1.3). The name suggests another useful way of thinking of the phenomenon; the permeability is a measure of the relative ease with which magnetic lines of force pass through the material. Permeability, like **dielectric constant** (3.4.2), is always measured relative to a vacuum (or in practice air, which is nearly the same); the permeability of soft iron is several thousand times that of air. Thus, in the design of magnetic apparatus, the lines of force are always made to 'flow' in a path situated as far as possible in a material of high permeability. This path is called the **magnetic circuit**, by analogy with an electric circuit. The high efficiency of the permanent magnet in Fig. 6.4 (6.1.3) is thus explained by the fact that its magnetic circuit consists of high permeability material, except for the small air-gap.

The outstanding advantage of the **electromagnet** over the permanent magnet is that the former can be switched on and off at will. This property is used in many pieces of electrical equipment, but one of great importance in radiology is called the **relay**. This appears in X-ray generators in a form called the **contactor**.

The principle of the relay is illustrated in Fig. 6.9. It is often required to switch on and off, from a distance, a large current from a large source of electrical power B_1, to an apparatus X. This, for example, is a common problem in X-ray apparatus design. If this were done directly, as in Fig. 6.9a, long heavy cables would be required to carry the large current without excessive voltage drop (5.3.2). Moreover, the switch Sw_1 would need to be a large and cumbersome type which could not be operated quickly and conveniently (e.g. as is required of the fluoroscopy switch on an X-ray set).

Instead, the arrangement of Fig. 6.9b is employed. In this case, the switch that carries the heavy current to the apparatus X takes the form of two large contacts C_1 and C_2. The upper contact C_1 is mounted on a strip of magnetic material A (known as the **armature**) pivoted (P) at one end and normally held back against a stop S by a spring Sp. The circuit is thus normally broken and the apparatus is switched off. Close to the armature A is fixed an electromagnet that is energized by the passage of a relatively small current through its coil from source B_2, controlled by the switch Sw_2. Although the cables to X must still be heavy, the whole solenoid and contact arrangement, which constitutes the relay or contactor, can be situated adjacent to X so that

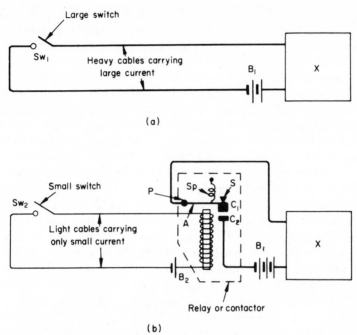

Fig. 6.9 The principle of a relay or contactor.

long heavy cables are not necessary. Of course, the cables to the switch Sw_2 must be long, but because the electromagnet needs only a small current from B_2 to operate it, the cables can be thin and light and Sw_2 can be small and easily controlled. We shall encounter the practical use of the relay or contactor again in the section on X-ray control circuits (13.4.1).

6.3. FURTHER APPLICATIONS OF THE MAGNETIC EFFECT

6.3.1 Electrical measuring instruments: the moving-iron meter. In section 6.2.1 we briefly mentioned the three basic requirements of an electrical measuring instrument:

   (i) an indicating device,
  (ii) an arrangement to produce a *force* proportional to the *current,* and
 (iii) a calibrated restoring force.

These basic requirements are not peculiar to electrical measuring instruments alone; a mechanical analogy that might assist the reader's understanding is found in the old-fashioned butcher's spring balance (1.3.3). In this, the meat, whose 'mass' must be measured, and which produces a downward force proportional to its mass, is hung on a hook attached to the lower end of a spring. The spring, naturally, is stretched; in the process, it exerts an upward force (the *restoring* force) on the meat and the two forces come to balance or **equilibrium**. The amount of stretch of the spring is then indicated on a scale by a pointer attached to the lower end of the spring. It is clear that the accuracy of the weighing depends critically on the spring being of the correct thickness and elasticity; if it is damaged by heating or excessive stretching an incorrect weighing will be obtained.

The most widely used 'meter' for measuring electric current is called the **moving-coil meter**. However, this is rather complex, and will be described in section 6.3.2. To illustrate the above principles we shall now describe the operation of one form of **moving-iron meter**; this type is less used but has certain advantages, one of which is that its operation is very simple.

Figure 6.10 shows a simplified diagram of this type of meter. The current to be measured flows through a solenoid S; in doing so it magnetizes equally two soft-iron rods (seen end-on) M and F. F is fixed to the body of the meter while M is attached to the lower end of a pivoted pointer P which travels over a scale Sc. The pointer is fitted with a light-weight coiled restoring spring Sp.

Current passing through the solenoid produces a proportional magnetic field that magnetizes the soft-iron rods M and F equally and in the same direction. These temporary magnets therefore repel each other with a force that increases with the value of the current. The pointer is thus driven forward over the scale until the repulsive force originating with the current is

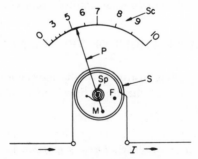

Fig. 6.10   A moving-iron meter.

balanced by the restoring force of the stretched spring. The scale is **calibrated** by passing known values of current through the coil and marking the position of the pointer for each value.

The moving-iron meter has the advantage that the direction of the pointer movement does not depend on the direction of the current, for if the current reverses, the directions of the magnetism of *both* M and F reverse together and they continue to repel each other. There are two disadvantages. First, the deflexion is not proportional to the current, so that the scale is not **linear**. Second, the pointer and the rod M are free to swing and in fact do so like a pendulum, many times, until the final reading of the pointer is attained. The instrument is said to be **undamped**. The latter disadvantage can be overcome with difficulty by providing friction devices containing air or oil; these control or **damp** the movement of the pointer without affecting its final position. We shall explain in sections 6.3.2 and 7.1.4 how the fundamental mode of operation of the moving-coil meter avoids these disadvantages.

6.3.2   The moving-coil meter is very widely used in electrical apparatus, including X-ray generators; it will repay detailed study for this reason and also for the way in which it illustrates fundamental laws.

Figure 6.11a shows a perspective sketch of a moving-coil meter. A 'horseshoe' magnet (Fig. 6.4, 6.1.3) has two pole-pieces N and S with a soft-iron cylinder C between them; only the pole-pieces of the magnet are shown in Fig. 6.11. The object of these particular pole-piece shapes is to produce a **radial magnetic field**, in which the lines of force in the air-gap between the pole-pieces and cylinder appear to radiate from the centre of the cylinder (Fig. 6.11b). Situated in the air-gap and pivoted so that it is free to turn is a rectangular coil of wire, usually supported on an insulated aluminium framework or 'former' (7.1.4). The coil is pivoted on two vertical spindles S supported in jewelled bearings B. Current is led into and out of the coil by two light-weight coiled springs Sp which also supply the restoring force for the meter (6.3.1). A pointer P is attached to the upper spindle and travels over a scale Sc.

The mode of operation of the moving-coil meter depends on the **motor effect** (6.2.2 and Fig. 6.6). It will be recalled that field, current and resultant force are all at right-angles, and in the meter the only parts of the coil which pass through the magnetic field at right-angles to it are those sides which are parallel to the axis of the cylinder. These are shown in Fig. 6.11b for clarity in the form of cross-sections of single conductors, i.e. the coil is shown as

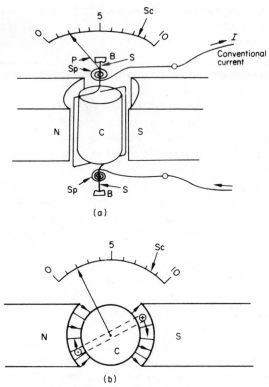

Fig. 6.11   A moving-coil meter.

having a single turn, though most meters have multi-turn coils. Now assume that a current is passed through the coil in the direction (conventional current) shown in the Figure. Fleming's left-hand rule (6.2.2) tells us that forces will be exerted on both conductors in the directions shown in Fig. 6.11b, i.e. in opposition to the restoring force of the springs. These will have the effect of driving the pointer forward over the scale to its equilibrium position. The purpose of the radial shape of the magnetic field can now be seen. At every position of the coil (between the normal limits of its travel) the resultant force will be at right-angles to the magnetic field, i.e. at right-angles to the radius or along the tangent to the circular path of the coil. The effect of this, combined with the **uniformity** of the magnetic field, ensures that the angle through which the pointer is deflected is exactly proportional to the current. The meter scale is thus linear, an advantage over the moving-iron meter. Although the meter, as described, is undamped, and would therefore suffer from troublesome oscillations about its final position (6.3.1), it will be shown in section 7.1.4 how a very satisfactory form of damping can easily be applied.

The moving-coil meter normally measures *current,* and is usually designed to give full-scale deflexion (f.s.d.) for milliamps or even microamps. It can then be converted (4.3.2) to read amps, volts, kilovolts, etc., as required. A change in its mechanical design will enable it to measure **charge** directly; this is the basis of the well-known **mAs meter** (13.5.2). Unlike the moving-iron meter, when the current is reversed its deflexion reverses (Fleming's left-hand rule — work it out yourself!). When this is a disadvantage, e.g. in the measurement of **alternating** current (8.1) a small circuit addition, called a **rectifier,** overcomes the difficulty (12.4.1). The moving-coil meter is such a versatile instrument that it finds many applications in radiological apparatus.

6.3.3 Electric motors. The **motor effect** was described earlier (6.2.1) as consisting fundamentally of the conversion of *electrical* energy to *mechanical* energy. This conversion process is very important in many applications; mechanical energy is often most useful in the form of **rotary** motion. A machine for converting electrical energy to rotary mechanical energy is called an **electric motor.** We shall not discuss electric motors in detail, because the simplest ones are not used in radiology, and the operation of the more advanced ones is too complex for this book. However, the following explanation will illustrate the general principles.

The moving-coil meter of Fig. 6.11 (6.3.2) is a kind of electric motor in that it produces limited rotary motion. If the pole-pieces could be made to enclose the core C almost completely (Fig. 6.11b), and if the whole apparatus were made bigger and stronger, with a strong shaft instead of pivots and jewelled bearings, the arrangement could produce appreciable mechanical energy. But this would be available for only one half-revolution of the coil, because when the coil sides reached the opposite part of the magnetic air-gap the resultant force would reverse (Fleming's left hand rule), and the coil would not continue to rotate indefinitely. To produce continuous rotation from a steady current a device is required which will reverse the current through the coil twice in each revolution; this is mounted on the motor shaft and is called a **commutator.** Such motors are not used in radiology.

A more advanced type of electric motor is used in certain types of X-ray tube. Because its functioning depends on **electromagnetic induction** (Chapter 7), it is called an **induction motor**. Its operation will be explained later (11.3.4).

7 Electromagnetic induction

7.1 ELECTROMAGNETIC INDUCTION

7.1.1 The converse of the motor effect. It was shown (6.2.1) that when electrons flow along a conductor, a magnetic field is produced. This magnetic field interacts with other magnetic fields which may be present (for example, if a permanent magnet is nearby) to produce a *force* between the two fields. The force may produce *movement* if one or both of the parts concerned are free to move. This is called the **motor effect.**

If an attempt is made to reverse the motor effect, for example by positioning a *stationary* magnet inside a coil (solenoid), no electron flow results through the coil. It appears that 'magnetism' cannot be converted into 'electricity' under these conditions.

If, instead, the magnet is *moved* inside the coil, an electron flow or *current* will result (if the coil is part of a complete circuit) and will continue so long as the magnet is moving. It seems that it is not the magnetism but something associated with the *movement* that is converted into the current; the magnetic field seems to act as a link between the *mechanical* movement and the *electron* movement. This phenomenon is called **electromagnetic induction.**

If the experiment is repeated with the coil not connected to anything, of course no current will flow at any time. However, suitable test instruments can show that during the movement of the magnet an *electromotive force (e.m.f.)* (5.3.2) is generated in the coil. This e.m.f. will naturally cause a proportional current to flow when there is a complete circuit. We say that the movement of the magnet causes the appearance of an **induced e.m.f.** in the coil. The term 'induced current' which is sometimes used is thus a misnomer, as it is the *e.m.f.* that is induced, the current merely resulting from the e.m.f. in accordance with Ohm's law (4.2.2).

Electromagnetic induction may be regarded in terms of energy conversion. The *mechanical* movement, being caused by a *force,* results in the expenditure of *mechanical* energy. The resulting induced e.m.f. and current together are equivalent to electrical energy (5.1.1, 5.1.2). Thus electromagnetic induction is a phenomenon by which *mechanical energy* can be converted into *electrical energy.* It is the fundamental process in every power station, in which mechanical energy derived from a steam turbine is converted into electrical energy by a machine called an alternator (8.1.2).

Electromagnetic induction is a very important phenomenon, for many reasons. Two of its principal applications are concerned with **alternating current** (8.1), which is the type of current generated in our power stations and supplied to hospitals, factories, houses, etc. Electromagnetic induction makes it possible to *generate* alternating current easily and relatively cheaply, as well as to *transform* it easily from one voltage to another. The two devices concerned in these processes are the **alternator** and the **transformer** (8.1.2 and 8.2.1).

7.1.2 The first law of electromagnetic induction. Figure 7.1 shows a coil connected to a centre-zero milliammeter. The latter, by indicating current flow, demonstrates the presence or otherwise of an induced e.m.f. in the coil. The permanent magnet shown can be moved in different ways relative to the

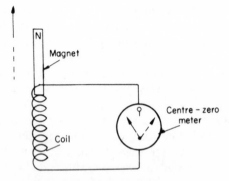

Fig. 7.1   Electromagnetic induction; a moving magnet.

coil. The following simple experiments with this apparatus demonstrate important phenomena, all of which are summarized in the **first law of electromagnetic induction**.

If the magnet is *thrust into* the coil, an e.m.f. will be induced in *one* direction. If the magnet is *withdrawn from* the coil, an e.m.f. will be induced in the *opposite* direction. Therefore reversal of the movement results in a reversal of the e.m.f. Similarly, reversal of the e.m.f. can be produced either by reversing the magnet or by reversing the direction of winding of the coil. If *two* of these three factors are reversed *simultaneously* the e.m.f. will remain in the same direction.

If the magnet is moved *more quickly* relative to the coil, a *larger* e.m.f. will be induced during the movement. Similarly, a *stronger* magnet will result in a *larger* e.m.f.

The primary effect of moving the magnet can be thought of in two alternative ways. In the first, the movement is imagined to result in the conductor 'cutting through' the lines of force (6.1.3) from the magnet. In the second, the magnet is regarded as producing a 'flow' or **flux** of lines of force through the circuit formed by the conductor; movement of the magnet then produces a **change of magnetic flux** linked with the circuit. These two ways of regarding the phenomenon are equivalent; sometimes one, sometimes the other is more convenient.

Hence the following may be deduced:

(i) if the magnet does not move relative to the coil, no e.m.f. is induced, and

(ii) if there *is* relative movement, an e.m.f. is induced which is proportional to the rate of cutting of lines of force, or in other words to the rate of change of magnetic flux linked with the circuit.

If the same experiment is tried with a coil having *more turns* of wire, the induced e.m.f. will be *larger*. In fact the induced e.m.f. is proportional to the total area of the circuit, i.e. the cross-sectional area of the coil multiplied by the number of turns.

The results of all the above experiments are summarized in **the first law of electromagnetic induction**, which states:

DEFINITION **A change of magnetic flux through a circuit causes an e.m.f. to be induced in the circuit. The induced e.m.f. is proportional to the rate of change of magnetic flux and to the area of the circuit.**

7.1.3 Moving conductors; Fleming's right-hand rule. The first law of electromagnetic induction may be demonstrated in a slightly different way. An e.m.f. is induced whenever there is *relative* movement between the coil and the magnet; thus the magnet may be held still and the coil moved instead. From this it is just a short step to a straight conductor moving in a magnetic field of parallel lines of force. This type of arrangement was used to demonstrate the motor effect (6.2.2) and is shown again in Fig. 7.2.

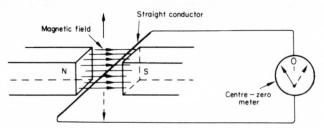

Fig. 7.2   Electromagnetic induction; a moving conductor.

When the conductor is moved *at right-angles* to the magnetic field, it *cuts* the magnetic lines of force, and an e.m.f., which obeys the first law of electromagnetic induction, is induced in the conductor. This way of demonstrating the law makes it very easy to predict the *direction* of the induced e.m.f. The directions of the movement, of the magnetic field and of the e.m.f. (and therefore of the conventional current) are all at right-angles to each other.

The direction of the e.m.f. may then be predicted using **Fleming's right-hand rule** (compare **Fleming's left-hand rule** for the *motor effect*, 6.2.2). As before,

> the thuMb denotes the Movement,
> the First finger denotes the magnetic Field, and
> the sEcond finger denotes the E.m.f.

Figure 7.3a shows the directions of the thumb and fingers, and Fig. 7.3b two examples of the use of the rule (compare with Fig. 6.6, 6.2.2). Note that it is the direction of the *conventional* current flow which is the same as the direction of the e.m.f. indicated by the rule.

This particular example of electromagnetic induction, viz. a conductor moving in a stationary magnetic field, is important because it forms the basis of the alternator (8.1.2).

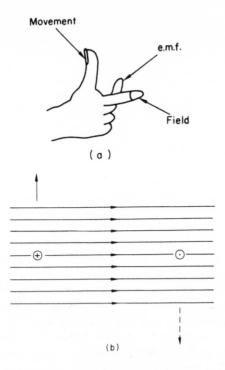

Movement

e.m.f.

Field

(a)

(b)

Fig. 7.3   Fleming's right-hand rule.

It may appear at this stage that perhaps it is not so important to be able to predict the direction of an induced e.m.f. However, this direction *in relation to other factors* is extremely important, and in section 7.1.4 it will be shown how a combination of Fleming's left- and right-hand rules lead directly to the **second law of electromagnetic induction.**

7.1.4   The second law of electromagnetic induction, or Lenz's law. Figure 7.4 shows a conductor lying at right-angles to a magnetic field. The ends of the conductor are imagined to be connected to a closed circuit which will permit the flow of current. If the conductor is moved, for example in an upward direction, Fleming's right-hand rule (7.1.3) tells us that an e.m.f. will be induced which in turn will drive a current through the conductor into the plane of the paper.

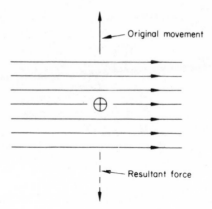

Fig. 7.4   Lenz's law.

However, there is now a current flowing through a conductor in a magnetic field, and a force will therefore be exerted on the conductor. The direction of this resultant force can be worked out from Fleming's left-hand rule (6.2.2); it is found to be in a downward direction. If the original movement had been downwards, the current would flow out of the paper, and the resulting force would be upwards. **The resulting force is always in the direction opposite to the original movement**; it only exists, however, *during* the movement. When there is *no movement,* there is *no induced e.m.f.* and *no resulting force.*

This is an example of a very widespread principle in science; in this case it is called *the second law of electromagnetic induction* or *Lenz's law,* which in more general form states:

DEFINITION   **An induced e.m.f. is in such a direction that it opposes the change which causes it.**

The reason for this more general wording is that induced e.m.f.s may result from factors other than movement (7.2.1); Lenz's law applies to all possible situations.

An important example of the application of Lenz's law in practice is that of the **electromagnetic damping of moving-coil meters.** The troublesome oscillation described in sections 6.3.1 and 6.3.2 can be completely eliminated by winding the coil on a framework or **former** made of aluminium, which is a good electrical conductor as well as being light in weight.

When the coil is deflected as a result of the current to be measured flowing through it, the sides of the aluminium former cut the radial lines of force in the magnetic field of the meter (Fig. 6.11, 6.3.2). An e.m.f. is induced in the aluminium, resulting in a flow of current because the aluminium acts as a single closed-circuit conductor. The current produces a magnetic field that interacts with the field of the meter magnet to produce a force opposing the movement, according to Lenz's law. The force exists *only while the coil is moving;* when the coil has settled down to its final correct position there is no opposing force. Therefore the damping does not affect the final position of the pointer, only the manner in which it reaches that position.

7.2 MUTUAL INDUCTION AND SELF-INDUCTION

7.2.1 Mutual induction. An e.m.f. is induced in a coil whenever there is a change of magnetic flux through the coil. This was produced in section 7.1.1 by moving a permanent magnet. The same effect can of course be obtained by replacing the *permanent magnet* by an *electromagnet,* consisting of another coil of wire wound around an iron core and connected to a battery. Current from the battery flows through the coil producing a magnetic field; this magnetizes the iron core causing a great increase in the total magnetic flux (6.2.3). If the electromagnet is now moved into and out of the first coil, exactly the same effects as were described in section 7.1.1 will be produced.

Using the electromagnet, it is not necessary to have mechanical movement for an e.m.f. to be induced. If it is remembered that the sole requirement is a *change of magnetic flux,* it will be clear that an induced e.m.f. will result if the current through the electromagnet is *switched on and off.* Switching the current *on* is equivalent to *inserting* the permanent magnet, while switching the current *off* is equivalent to *removing* it. This particular form of electromagnetic induction is called **mutual induction,** for reasons to be described later. The two laws of electromagnetic induction apply equally to this form as to the form previously described (7.1.1).

7.2.2 The induction coil; transformers. Mutual induction is demonstrated by the apparatus shown in Fig. 7.5. The part enclosed in the dashed line, if made as a complete piece of apparatus, is called an **induction coil.** The primary and secondary coils are usually wound one on top of the other, both being around the iron core, but in the conventional symbol the three components are shown separately for clarity.

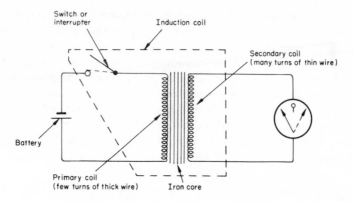

Fig. 7.5   The induction coil; mutual induction.

If an e.m.f. is induced by movement of a *permanent magnet* (7.1.1), *mechanical* energy is transformed into *electrical* energy. In *mutual* induction, the *electrical* energy from the battery in the *primary* circuit is transformed into *electrical* energy in the *secondary* circuit. The magnetic field, as before, acts as a link between the two circuits. In fact, the link provided by the magnetic field can act in *both* directions; as well as effects in the secondary coil being produced by causes in the primary coil, effects in the primary coil can be produced by causes in the secondary coil. This is the reason for the term *mutual.*

The induction coil has few uses at the present day, but formerly it had many uses, one of which was to produce the very high voltages necessary to operate an X-ray tube (11.2.2). In such applications a continuous transfer of electrical energy from primary to secondary circuits was necessary; this was achieved by replacing the switch in series with the battery by an interrupter which continually switched the current on and off. Each 'on' and each 'off' caused a change of magnetic flux and hence an induced e.m.f. in the secondary coil.

The most important property of the induction coil is its ability to produce *high* voltages from *low.* This it is able to do by a suitable ratio of the numbers of turns on the primary and secondary coils. Thus if the *primary* has a *small* number of turns of *thick* wire, and the *secondary* a *large* number of turns of *thin* wire, a *small* voltage applied to the *primary* will induce a *large* e.m.f. in the *secondary.* This could be made as much as 50 to 70 kV, i.e. enough to operate an X-ray tube (11.2.2).

The induction coil is now obsolete for this purpose but is still in universal use as the ignition coil in petrol engines. For X-ray and other uses it has been replaced by the **transformer** used in conjunction with alternating current (8.2.1). The most important application of mutual induction is thus in the **alternating current transformer** which has the ability to transform one voltage into another cheaply and efficiently.

7.2.3 Self-induction; inductance. In the induction coil, the changing primary current causes a change in magnetic flux which induces an e.m.f. in the *secondary*. However, the magnetic flux passes through the primary coil as well, and an e.m.f. is therefore induced in the *primary* coil also. This process will occur even when no secondary coil is present; it is called **self-induction** (Fig. 7.6).

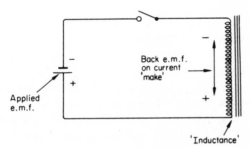

Fig. 7.6   Self-induction.

The special characteristics of self-induction result from the direction of the *induced* e.m.f. relative to that of the *applied* e.m.f. (i.e. that from the battery). Lenz's law states that the induced e.m.f. is in such a direction that it opposes the change that caused it. This change is the rising current produced by switching on the applied e.m.f. Hence the induced e.m.f. *opposes* the applied e.m.f. and is called a **back e.m.f.**

The back e.m.f. therefore tends to oppose the *rise* of current that caused it; if the current *falls*, the back e.m.f. reverses and tends to oppose the *fall* of current. Hence self-induction always tends to oppose any *change* in the current flowing through the coil. This is an important effect in alternating current circuits because alternating currents are continuously changing; the coil is said to have the property of **self-inductance** or simply **inductance**.

8 Alternating currents and transformers

8.1 ALTERNATING CURRENTS

8.1.1 The advantages of alternating currents. So far, in this book, the electric currents discussed have been thought of as constant and flowing in one direction only. These are called **direct currents**. For purposes of example and illustration they have been produced by a **cell** or a **battery** (of cells) (4.1.2) both of which produce a constant **e.m.f.** (4.1.3, 5.3.2) by the conversion of chemical energy into electrical energy.

Such cells or batteries are useful for producing relatively small amounts of electrical energy for portable use, e.g. in torches, transistor radios and motor cars. However, the cost of the energy they produce is relatively very high, and for large amounts of energy such as are required in factories, hospitals, etc., the cost of batteries would be prohibitive. For example, the cost of one B. of T. Unit (5.1.2) from the Electricity Board supply is about 3p, whereas from batteries it might be as much as £15 or more.

Most of the electrical energy in use today is **generated** in power stations, making use of the phenomenon of electromagnetic induction (7.1.1). Steam is first produced, by burning coal or oil or from nuclear power; the steam drives a turbine which itself drives an electrical machine called a **generator**. The generator produces electrical energy. Thus the sequence of energy changes is

$$\text{chemical (coal or oil)} \xrightarrow{\text{burning}} \text{heat} \xrightarrow{\text{steam turbine}} \text{mechanical} \xrightarrow{\text{generator}} \text{electrical.}$$

In districts such as the north of Scotland, where water power abounds, the water itself drives water turbines. Although the average coal-burning power station, for example, is only about 28% efficient, i.e. 28% of the energy in the coal is eventually converted into electrical energy, the final cost is relatively low.

For the sake of efficiency, it is best to use the simplest form of electrical generator. As will be seen later (8.1.2), this is called an **alternator**; it generates a type of current called **alternating current** which reverses periodically at a rate of 100 reversals per second (in the U.K.). This peculiar and complex form of current seems at first sight to be far less desirable than the simple direct current. However, although it is possible to generate direct current with a machine, the latter is more complex and difficult to maintain than an alternator. Besides, alternating current has another very important advantage, as follows.

Electrical energy is supplied to houses, factories and hospitals for use mainly at a voltage (5.3.2) of about 240 volts. Suppose we wish to have either a lower or a higher voltage for some special purpose (this requirement is very common). If the supply is direct current, we should need to use very inefficient and perhaps expensive and elaborate apparatus to 'transform' the voltage. If the supply is alternating current, however, the change of voltage can be achieved very simply, cheaply and with high efficiency by a device called a **transformer** (8.2.1).

To summarize, therefore: **alternating current** is widely used today because

(i) it is easily and cheaply **generated**, and
(ii) it is easily and cheaply **transformed**, from one voltage to another,

using devices called **alternators** and **transformers** respectively.

8.1.2 The generation of alternating currents; the alternator.

The alternator makes use of the principle of electromagnetic induction (Chapter 7). The most important feature of its design is a means of causing a conductor to 'cut' magnetic lines of force continuously. This is achieved by the arrangement shown in perspective in Fig. 8.1a. A permanent magnet such as that in Fig. 6.4 (6.1.3) produces a parallel magnetic field (for clarity not shown) between pole-pieces N and S. A coil or **armature** A rotates between the pole-pieces; it is shown with a single turn of wire but in practice many turns are required to produce the desired performance. The rotation is continuous, being produced by mounting the coil on a shaft Sh, driven round by a pulley P and belt from

93

some kind of engine. (In practice, direct drive or gears would be used.) In cutting the lines of a magnetic force, an e.m.f. is induced (7.1.3) in the coil. To make use of the e.m.f. and to cause it to produce a current, connexions must be made to the ends of the wire forming the coil. These connexions

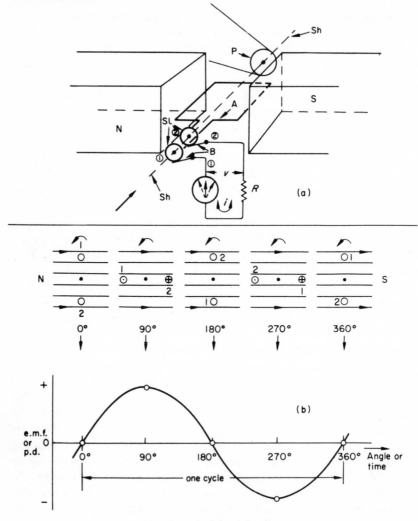

Fig. 8.1　The principle of the alternator.

cannot be made directly, because they would rapidly become twisted as a result of the continuous rotation. This difficulty is avoid by using two bands of copper called **slip-rings** Sl, mounted on the shaft but insulated from it, each connected to one end of the coil. Rubbing connexion is made to the slip-rings via two hard carbon **brushes** B which are connected to the external circuit. In the arrangement shown, the induced e.m.f. drives a current through a resistance R which may represent a lamp, heater, etc.

It is important to know what kind of e.m.f. is induced in the coil. This may be deduced qualitatively by considering various rotational positions of the armature A. The upper half of Fig. 8.1b shows a series of views of the alternator looking along the shaft in the direction of the large arrow in Fig. 8.1a. The parallel lines represent the magnetic field, and the five small drawings show the five positions of the armature at 90-degree intervals in turning through a complete revolution, or **cycle**. Only those parts of the armature which are at right-angles to the magnetic field are shown, because only these have any e.m.f. induced in them. Each side of the armature is identified (by 1, 2) to correspond to the slip-ring to which it is connected (Fig. 8.1a).

It is not possible, of course, to *calculate* the e.m.f. induced at each of the five positions shown, but an *estimate* can be made of its magnitude and direction, as follows. The e.m.f. is proportional to the rate of change of magnetic flux or to the rate at which the magnetic lines are 'cut' (first law of electromagnetic induction, 7.1.2). It is clear that at the $0°$, $180°$ and $360°$ positions the lines are not being cut at all, because the conductors are momentarily travelling *parallel* to the lines. At these positions the e.m.f. is therefore zero. At the $90°$ and $270°$ positions the conductors are momentarily travelling *at right-angles* to the lines and are therefore cutting them fastest; the e.m.f. at these positions is therefore a maximum. But if we apply Fleming's right-hand rule (7.1.3) we discover that at position $270°$ the e.m.f. is *opposite* to that at position $90°$; conductor no. 1 is positive at $90°$ but negative at $270°$.

Now let us incorporate these findings in a graph, as in the lower half of Fig. 8.1b. In this graph is plotted the e.m.f. in the coil, or the p.d. across the brushes, against the angle of rotation of the armature. The points on the graph illustrate the values of e.m.f. deduced above. It can be seen that the e.m.f. starts at zero, reaches a maximum or **peak** in the positive direction, drops to zero, goes to a negative peak then returns again to zero. The whole series of operations is called one **cycle**. Obviously we have produced an e.m.f.

which periodically reverses, for continuous rotation of the armature would produce a succession of such cycles. This is called an **alternating e.m.f.**, and the current which it would produce in the resistance R is an **alternating current.**

We cannot deduce by simple reasoning what would be the value of the e.m.f. *between* the positions discussed above. It can only be stated that the rate of change of magnetic flux is proportional to the trigonometrical sine of the angle θ* of rotation, hence the e.m.f. is proportional to $\sin \theta$. The shape of this curve can be obtained from trigonometrical tables, and has been drawn in the graph. The whole graph therefore represents one cycle of a particular kind of alternating quantity known as a **sinusoidal** quantity, i.e. one which varies as the sine of an angle. A sinusoidal e.m.f. is the simplest kind (strange to say!) of alternating e.m.f., and is the kind supplied by the country's power network.

The apparatus shown is useful only for demonstration. A practical alternator as seen in a power station is a huge machine looking quite unlike Fig. 8.1a. However, the principles on which it operates are identical to the above.

8.1.3 Frequency and waveform. In the continuous operation of the demonstration alternator of Fig. 8.1 (8.1.2), one **cycle** follows another at a constant rate, this rate being known as the **frequency** of the alternating quantity. For example, if the armature rotates 50 times per second, the frequency so produced will be 50 **cycles per second**, abbreviated to 50 c/s. Sometimes the 'cycle-per-second' is called the **hertz** (abbreviated to Hz); the above frequency would be called 50 Hz. This is in fact the frequency of the mains supply in the United Kingdom and in Europe; in some other countries, e.g. the U.S.A., the frequency is 60 c/s or 60 Hz.

The graph of the single cycle shown in Fig. 8.1b can apply to e.m.f., p.d., or current, and its shape is called the **waveform** of the alternating quantity. In this case the waveform is **sinusoidal** (8.1.2), and is characteristic of the mains supply. However, many other types of waveform are encountered and are useful in different applications. The term 'waveform' is in one sense unfortunate, as it conjures up impressions of waves in the sea and may lead to misconceptions about the true nature of alternating currents. In other words, there is no 'wave' (in the sense of a water wave) in the conductor; the

* θ is the Greek small letter theta.

waveform is merely a graphical method of showing how the e.m.f., current, etc. varies with angle or with time. It shows that the e.m.f. starts at zero, increases in one direction to a maximum, decreases to zero, then reverses and increases to a maximum in the other direction, finally decreasing once more to zero.

8.1.4 Peak and effective (r.m.s.) values. There is no problem about assigning a value to the e.m.f., current, etc. of an unvarying direct current, because the quantities concerned are constant. Figure 8.2a, for example, shows p.d. and current in a d.c. circuit, the e.m.f. being produced by a cell. One could say for example that the p.d. (V) is 10 volts and the current (I) is 2 amperes.

Figure 8.2b shows the corresponding alternating current circuit, with the conventional symbol for an alternator. The latter also serves as the symbol for any source of alternating e.m.f. In this circuit the e.m.f., p.d. and current are continuously varying; how then can we assign a significant value to them? There are several possibilities (Fig. 8.3a), as follows.

(i) The **instantaneous** values, v and i. These merely give the value at any specified *instant* in time, and are of limited application in normal use.

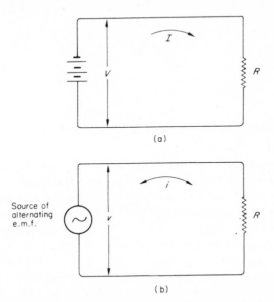

(a)

(b)

Fig. 8.2   Direct current and alternating current circuits.

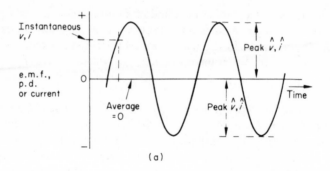

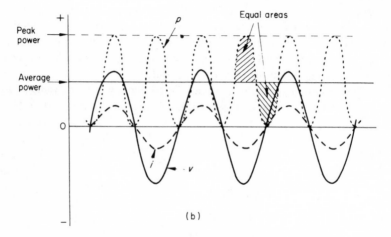

Fig. 8.3   Instantaneous, peak and effective (r.m.s.) values.

(ii) The **peak** values, $\hat{v}$ and $\hat{i}$, which are of course the *maxima* of the instantaneous values. The peak values, although useful in some circumstances, are of limited application because they occur only for two instants in each cycle. For the remainder of the time the e.m.f., etc., has lower values.

(iii) It may seem logical to overcome the latter disadvantage by using the **average** or **mean** values. However, because alternating quantities are equally positive and negative, the average value is zero. In fact, this serves as a definition of an alternating quantity: one whose average value is zero.

(iv) The most useful measure of an alternating quantity for normal use would be a *kind* of average (which was not equal to zero) which would be representative of the *effect* of the alternating current, perhaps by comparison with a direct current in terms of a criterion such as power production or heating effect. This value is called the **effective** e.m.f., p.d. or current.

DEFINITION   **The effective value of an alternating current is that value of direct current which has the same average heating effect as the alternating current.**

The question we have to answer is: how is the *effective* value related to the *peak* value? To answer this, we must discuss further the behaviour of alternating current in a circuit such as in Fig. 8.2b. We know that the flow of *direct* current through a resistance produces an amount of heat per second proportional to I^2R (5.2.1). What happens in the case of *alternating* current? Common sense tells us that the reversal of current makes no difference because heat is produced regardless of the direction of current flow. The rate of heat production will of course vary during the cycle; in fact at any instant it will be proportional to i^2R where i is the instantaneous current.

The above is a commonsense view; we can show that direction of current is unimportant by the following reasoning. The alternator (Fig. 8.2b) produces an alternating e.m.f. which drives a current i through the resistance R, resulting in a p.d. across R (5.3.2). Now Ohm's Law applies equally to alternating current as to direct, hence the instantaneous current i is at all times proportional to the instantaneous p.d. v. This is shown in Fig. 8.3b where the waveforms of v (full line) and i (dashed line) are seen to be 'in step' or **in phase**. This means that they are zero together and reach their positive peaks and negative peaks together. Now the instantaneous power p, and therefore the instantaneous rate of heat production, is vi. The waveform of this product, viz. the **power curve**, is shown dotted. When both v and i are positive, the power is clearly positive. When the current reverses, both v and i are negative, but *their product is positive* (negative x negative = positive!). Hence each **half-cycle** of current, no matter what its direction, produces heat.

The same diagram can be used now to deduce the relation between **effective** and **peak** values. From the above, the instantaneous power is given by $p = vi = i^2R$ (5.2.1). Hence the peak power is given by

$$\hat{p} = \hat{i}^2R. \qquad\qquad \text{Eq. 8.1}$$

The effective value however is defined above in terms of *average* power or heating effect; it can be guessed from the symmetry of the power curve as shown in Fig. 8.3b (and it can also be proved mathematically) that the average power P is one half of the peak power. Hence, from Eq. 8.1

$$P = \tfrac{1}{2}\hat{p} = \tfrac{1}{2}\hat{\imath}^2 R \qquad \text{Eq. 8.2}$$

Now suppose I stands for the *effective* value of the current (as well as for the value of the direct current to which it is equivalent), then from Eq. 8.2:

$$P = I^2 R = \tfrac{1}{2}\hat{\imath}^2 R$$

Therefore $I^2 = \tfrac{1}{2}\hat{\imath}^2$

and $I = \sqrt{(\tfrac{1}{2})} \, . \, \hat{\imath} = \hat{\imath}/\sqrt{2}.$ Eq. 8.3

Similarly it can be shown that

$$V = \hat{v}/\sqrt{2}. \qquad \text{Eq. 8.4}$$

It is sufficiently accurate to take $\sqrt{2} = 1\cdot4$ and $1/\sqrt{2} = 0\cdot7$. Hence the relations between *effective* values and *peak* values can be summarized in the general statements:

$$\text{peak} \begin{vmatrix} \text{voltage or} \\ \text{current} \end{vmatrix} = \text{effective (r.m.s.)} \begin{vmatrix} \text{voltage or} \\ \text{current} \end{vmatrix} \times 1\cdot4 \qquad \text{Eq. 8.5}$$

and

$$\text{effective (r.m.s.)} \begin{vmatrix} \text{voltage or} \\ \text{current} \end{vmatrix} = \text{peak} \begin{vmatrix} \text{voltage or} \\ \text{current} \end{vmatrix} \times 0\cdot7 \qquad \text{Eq. 8.6}$$

The abbreviation **r.m.s.** stands for **root-mean-square** and is a reference to the form of the mathematical derivation of Eqs. 8.5 and 8.6. It is commonly used by engineers as equivalent to 'effective'; either term is acceptable.

In speaking of alternating supplies, if only a voltage or current is specified without qualification, it refers to effective or r.m.s. values. For example, the 240-volt alternating supplies in our homes are 240 volts *effective* or *r.m.s.* Numerical example: What is the peak voltage of the 240-volt mains supply?

$$\text{Peak voltage} = \text{effective } V \times 1\cdot4$$
$$= 240 \times 1\cdot4 = 336 \text{ volts.}$$

8.1.5   A classification of types of current, etc. We have already met direct and alternating currents, and later will encounter other types. The following is a broad classification of the main types of current used in radiology:

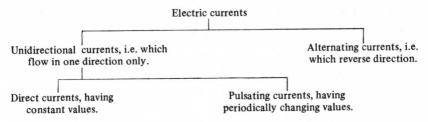

Typical waveforms of these currents are shown in Fig. 8.4.

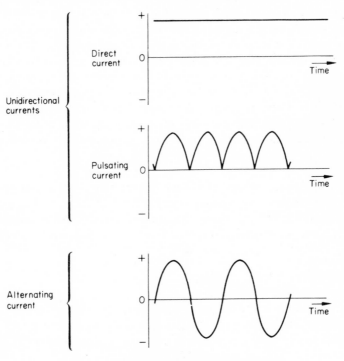

Fig. 8.4   Unidirectional and alternating currents.

8.2 TRANSFORMERS: THEORY

8.2.1   The induction coil and the transformer. In section 7.2.2 we described how electrical energy can be transferred from one circuit (the **primary**) to another (the **secondary**) via a link consisting of a changing magnetic flux. The changing magnetic flux is produced by continual interruption of the primary current (from a battery); the whole arrangement is called an **induction coil** (Fig. 7.5, 7.2.2). One of the great advantages of the induction coil is its ability to produce a high voltage from a low voltage; it was formerly used for this purpose in X-ray generators and is still so used in motor-car engines for ignition purposes.

Although reasonably satisfactory in the latter application, the induction coil is relatively inefficient and unreliable for high power use. If it is recalled that the energy transfer takes place because of the *changing* magnetic flux, however, it will be obvious that if *alternating* current is passed through the primary coil, a constantly changing (alternating) magnetic flux will be produced in the iron core, and hence an alternating e.m.f. will be induced in the secondary coil. The interrupter is not now required, and the arrangement is an elementary example of an **alternating current transformer**. The transformer is a simple, cheap and efficient means of changing alternating voltages and currents from one value to another, and is one factor which has led to the universal adoption of alternating sources of electrical power (8.1.1).

8.2.2   Turns ratio, voltage ratio and current ratio. Let us consider the transformer shown symbolically in Fig. 8.5. Whatever the type of construction, the iron core passes through both primary and secondary coils. Therefore the same change of magnetic flux flows through every turn of wire,

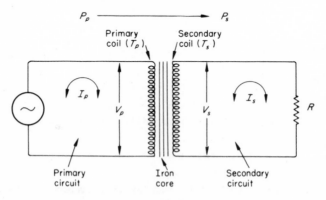

Fig. 8.5   The principle of the transformer.

whether on the primary or on the secondary; each *turn* of wire is therefore associated with the same e.m.f. or voltage. By suitably proportioning the relative numbers of turns on the primary and secondary, we can make the transformer either **step-up** or **step-down** in respect of voltage. Quantitatively, if T_p is the number of turns on the primary, and T_s is the number of turns on the secondary, then the turns ratio of the transformer is T_s/T_p, and this is equal to the **voltage ratio** V_s/V_p.

Hence
$$\frac{V_s}{V_p} = \frac{T_s}{T_p}.$$
Eq. 8.7

This is the basic voltage equation of the transformer. Thus, for example, if the secondary has five times as many turns as the primary, and the latter has 200 volts applied to it, the secondary e.m.f. will be 1 000 volts.

This example, which illustrates a **voltage step-up** transformer converting 200 volts to 1 000, appears to suggest that one is getting 'something for nothing'! That this is a mistaken idea can be revealed by applying to the transformer the law of conservation of energy (1.1.4). We can say that the power **output** (energy per unit time, 5.1.2) obtained from the secondary cannot exceed the power **input** to the primary. In fact, in an ideal transformer the two are equal.

Hence
$$P_s = P_p.$$
Eq. 8.8

To consider the power in the transformer circuits it is necessary to bring in the idea of the currents, for $P = VI$ (Eq. 5.6, 5.1.2). In Fig. 8.5, let us assume that the secondary circuit is broken (i.e. the resistance R is infinite). Then the secondary current I_s will be zero. In an ideal transformer the primary current I_p will also be zero. Now suppose R is made finite, i.e. some current is drawn from the secondary coil. Then the magnetic conditions of the transformer so adjust themselves that just enough primary current flows to supply the necessary input power. If the secondary current is further increased, the primary current increases to compensate. The transformer is thus a self-regulating device whose primary current always sets itself to just the correct value for equilibrium (in this case to satisfy Eq. 8.8).

Now
$$P_s = V_s I_s \quad \text{and} \quad P_p = V_p I_p;$$
then from Eq. 8.8
$$V_s I_s = V_p I_p$$
and, cross-dividing and applying Eq. 8.7

103

$$\frac{V_s}{V_p} = \frac{I_p}{I_s} = \frac{T_s}{T_p} \qquad\qquad \text{Eq. 8.9}$$

This equation shows that although the **voltage ratio** is *equal* to the **turns ratio**, the **current ratio** (in an ideal transformer) is equal to the *inverse* of the turns ratio. In other words, a *voltage* step-*up* transformer is a *current* step-*down* transformer.

In the above numerical example, if a current of 1 A flows in the secondary circuit, a current of 1 x 5 = 5 A would flow in the primary circuit. Thus the primary power would be 200 x 5 = 1 000 W and the secondary power 1 000 x 1 = 1 000 W.

This discussion assumes an *ideal* transformer, i.e. one without losses. We shall now see what happens when some of the energy or power is lost in flowing through the transformer.

8.2.3   Transformer efficiency. In the last section we assumed that we were dealing with a perfect transformer, viz. one which gave out from its secondary all the power that was put into the primary. Practical transformers are not ideal, and the **output** power is less than the **input** power. This aspect of the transformer may be measured in terms of its **efficiency**:

DEFINITION   **The efficiency of a transformer is equal to the output power divided by the input power (times 100 if required as a percentage).**

Hence

$$\text{Efficiency} = \frac{P_s}{P_p}, \qquad\qquad \text{Eq. 8.10a}$$

$$\text{Percentage efficiency} = \frac{P_s}{P_p} \times 100. \qquad\qquad \text{Eq. 8.10b}$$

It may be thought that this behaviour is contrary to the law of conservation of energy. What happens to the missing power (energy)? As in the case of the X-ray tube (5.2.1, 5.2.2) it is converted into heat because of certain non-ideal processes in the transformer. These processes will be described in the next section; it is important to realize now that the heat so produced has disadvantages similar to those for the X-ray tube, viz. (i) it represents a loss of energy, and (ii) it heats the transformer unnecessarily and perhaps harmfully.

Numerical example on transformer operation.

A transformer delivers an output of 100 kV (r.m.s.) and 2 mA (r.m.s.). What is the primary current if the primary voltage is 100 V and the efficiency 80%?

The secondary power

$$P_s = V_s I_s = 10^5 \times 2 \times 10^{-3} = 200 \text{ W}.$$

Now % efficiency $= \dfrac{P_s}{P_p} \times 100$ (Eq. 8.10b).

Hence $P_p = \dfrac{P_s \times 100}{\% \text{ efficiency}} = \dfrac{200 \times 100}{80} = 250 \text{ W}.$

Hence $I_p = \dfrac{P_p}{V_p} = \dfrac{250}{100} = 2 \cdot 5 \text{ A}.$

Transformer efficiency may also be considered in terms of internal resistance, regarding the transformer as a generator (5.3.2). The loss of power associated with lack of efficiency may be regarded as resulting from the flow of current through the internal resistance, with the consequent production of heat.

A numerical example similar to that relating to the car battery in section 5.3.2 can apply equally to a transformer. For example, with a given primary voltage, the secondary p.d. of an X-ray tube filament transformer might be 12 V (r.m.s.), when no current is drawn. If the tube filament is now switched on and draws 10 A, the secondary p.d. may fall to 11 V. This, by analogy with the previous example, would correspond to an effective internal resistance of 0·1 Ω.

This type of behaviour is often referred to as **transformer regulation**; a transformer with a *good* regulation is one which maintains its secondary p.d. more nearly constant (for a varying load current) than one with a *bad* regulation. The regulation is often described as a percentage change of secondary p.d. for a given load current. In the above example, the transformer regulation is $(1 \cdot 0/12) \times 100 = 8 \cdot 3\%$ for an output current of 10 A.

8.3 TRANSFORMERS: PRACTICAL ASPECTS

8.3.1 Transformer losses. The fact that the efficiency of a practical transformer is less than 100 % (8.2.3) implies that energy is lost, in this case in the form of heat, in the transformer. So that the energy loss may be reduced to a minimum it is important to understand what are the processes responsible for the loss. A transformer comprises two main types of component: the coils, made of copper, and the core, made of iron. Accordingly, it has been customary to divide the processes into *copper losses* and *iron losses*. However, this is not a good fundamental classification, and we shall consider them in two different categories: (i) 'current' losses, and (ii) 'hysteresis' losses.

(i) **Current losses** result from the fact that whenever a current I flows through a resistance R, an amount of power equal to I^2R watts is converted into heat (5.2.1). They are therefore sometimes known as 'I^2R' losses. They can arise in both (a) the copper coils and (b) the iron core.

(a) Current must flow in the coils of a transformer during its normal functioning. The coils, being non-ideal conductors, possess resistance. Therefore 'I^2R' losses arise, and electrical energy is converted into heat. To reduce this source of loss, the current cannot be reduced because the normal operation of the transformer would be affected. Instead, the resistance of the conductors must be minimized by using wire of low resistivity which is as thick as convenient (4.2.3). Copper is nearly the best conductor known, and the thickness of wire used must always be a compromise between cost, space and saving of power. In general, the larger the current in a coil, the thicker should be the wire. Thus, in a voltage step-up transformer, the primary current will be many times the secondary current, hence the primary coil will be wound with thicker wire than will the secondary.

(b) It is not at first apparent how current can flow in an iron core. However, Fig. 8.6a shows how the cross-section of the core may be regarded as consisting of concentric 'layers' of iron, each acting as a short-circuited single-turn secondary coil (in addition to the normal secondary). Hence e.m.f.s will be induced in the core; these will produce currents called **eddy currents** (by analogy with circular 'eddies' in a pool of water). The eddy currents, in flowing through the resistance of the core, will give rise to 'I^2R' losses, and the power for these is drawn from the primary circuit. Eddy currents are readily eliminated by making the iron core in the form of thin sheets of metal (Fig. 8.6b), each sheet being insulated from its neighbour by a thin layer of paper. The result, called a **laminated core**, is very nearly as effective magnetically as the solid core.

(ii) **Hysteresis losses** are totally different in nature from current losses. According to the molecular theory of magnetism (6.1.2), when a magnetic

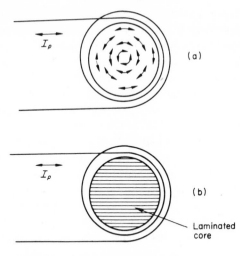

(a)

(b)

Laminated
core

Secondary coils not shown in either case

Fig. 8.6   The elimination of eddy currents.

material is magnetized, the 'molecular magnets' of which it is composed have to turn on their axes. It may be imagined that there exists some resistance to this turning − a kind of 'molecular friction' − and that this resistance will vary from one material to another.

When a magnetic material forms the core of a transformer, its magnetization is being reversed at twice the frequency of the current. At each reversal, energy is lost due to the 'molecular friction', and this lost energy must come from the source of the primary current. The 'molecular friction' effect is called **hysteresis**. It is readily reduced in practice by choosing for the core a suitable magnetic material, for example a steel alloy called 'Stalloy'.

8.3.2 Transformer construction. A practical transformer differs considerably from the cylindrical iron core and two simple coils of the induction coil (7.2.2). The main differences are as follows.

(i) Although there is seldom more than one primary coil or **winding**, a single primary winding very often serves more than one secondary winding. Thus, in a typical transformer used in control equipment, the primary might be designed for 200 V input, and there might be three secondary windings:

(a) 500 V output, low current, (b) 50 V output, low current, (c) 6 V output, high current. The primary would have wire of medium thickness, secondaries (a) and (b) thin wire, and secondary (c) thick wire.

(ii) The straight cylindrical iron core of the induction coil, which is also implied in the conventional symbol for the transformer, is very inefficient because so much of the **magnetic circuit** consists of air (6.2.3). Transformer cores are therefore always designed so that they form a **closed magnetic circuit**, which because of the high **permeability** (6.2.3) is very efficient. At the same time it is necessary to **laminate** the core to eliminate **eddy currents** (8.3.1), so that one type of small transformer (such as the example above) appears as in Fig. 8.7a. The core is made up of thin **T**- and **U**-laminations (Fig. 8.7b), which are stacked together, but insulated from each other, to form the required thickness. The primary and secondary windings are wound, one on top of another, on a bobbin which envelops the centre 'leg' of the core. The power efficiency (8.2.3) of a small transformer like this might be only 60%,

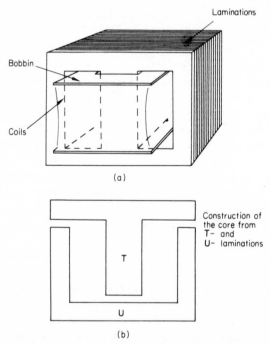

Fig. 8.7   The core and bobbin of a transformer.

but efficiency usually (and fortunately) increases with size, so that in large and expensive transformers it might exceed 95%.

(iii) Transformers designed for high voltages (in the region of kilovolts for X-ray work) call for special care in design. The secondary winding may produce as much as 100 kV (which is about 140 kV *peak,* or **kVp**), and the winding itself must be designed very carefully to avoid electrical breakdown due to ionization of the surrounding air (3.3.2). In addition, the p.d. between parts of the secondary and the primary may need to be even greater, so that the type of construction shown in Fig.8.7a may not be suitable. Instead, the cores of such transformers are in the shape of a laminated square and the coils between which the large p.d. appears are wound on highly insulating plastic bobbins mounted on opposite sides of the core (Fig. 8.8).

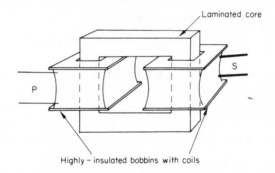

Fig. 8.8   A transformer designed for a high p.d. between windings.

Transformers for high voltages are usually enclosed in a metal tank filled with oil; the oil is a better insulator than air — it is said to have a higher **dielectric strength** (3.4.2). In addition, it also helps to carry heat away from the core and coils by **convection** (1.4.3); the heat is then removed from the metal case by *air* convection and by radiation (1.4.3).

8.3.3 Autotransformers. The most important advantage of the transformer is its ability to transform alternating voltages and currents up or down with relatively high efficiency and convenience. The type so far discussed is known as a **double-wound** transformer, that is its primary and secondary windings are completely separate electrically. Power is transferred from primary to secondary solely via the link provided by the ever-changing magnetic flux.

This electrical **isolation** between the windings is very often an essential feature; in other situations, however, it is not required. In these cases, another type called an **autotransformer** (often abbreviated to 'auto') may be used (Fig. 8.9). In this type, not only are the primary and secondary connected

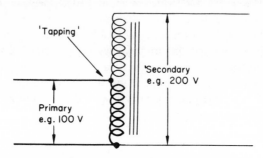

Fig. 8.9   The principle of the autotransformer.

together, but part of the winding actually serves a dual purpose. For example, the autotransformer in Fig. 8.9 has a 2 : 1 voltage step-up ratio. The transformer consists of a single winding on an iron core; the winding has one or more extra connexions, known as tapping points or **tappings**. In this example the tapping is to the centre point of the winding to give the ratio of 2 : 1.

The pattern of flow of currents in the autotransformer is complex; it is sufficient for our purpose merely to regard the lower half (in this example) of the winding as the primary, and the whole winding as the secondary. The auto will, of course, act equally well in reverse as a 1 : 2 voltage step-down transformer.

Autotransformers may have many tappings; they are widely used where moderate voltage ratios are required and electrical isolation between primary and secondary is not necessary. We shall see later that they occupy a very important place in most X-ray generator circuits (Chapter 13).

9 Electromagnetic radiation

9.1 ELECTROMAGNETIC WAVES

9.1.1 The production of electromagnetic radiation. An electric charge is surrounded by an electric field (3.1.2). If the charge is stationary, the electric field at any given point is constant. If the charge moves with uniform velocity (e.g. if there is a steady flow of electrons), a magnetic field also is observed (6.2.1). If the charge undergoes an *acceleration* (or a *deceleration,* which is of course a negative acceleration), both the magnetic and the electric fields at the point vary and they do so in the special way described in section 9.1.2. (An example of such a deceleration is found in a changing or alternating current, Chapter 8.) This combined variation of electric and magnetic fields results in loss of energy by the decelerating charge; the charge radiates (1.4.3) energy in a form known as **electromagnetic radiation** or **electromagnetic waves** (9.1.3). In fact, whenever an electrically charged particle undergoes a deceleration, energy is radiated by the particle in the form of electromagnetic radiation.

9.1.2 The characteristics of waves. If, at a given instant in time, the strength of the electric field (3.1.2) is measured at various distances from a source of the electromagnetic waves, it will be seen to vary in an alternating or *cyclic* manner with distance (Fig. 9.1). Such a cyclic variation, in which the electric field strength repeatedly changes in a gradual manner from a positive peak to a negative peak and back again, is described as **sinusoidal** because it can be

111

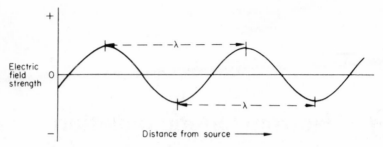

Fig. 9.1   The strength of the electric field plotted against distance from a source of electromagnetic radiation at a given instant in time, t_1.

represented mathematically by an equation involving the trigonometrical sine of an angle (see section 8.1.2 on alternating currents).

The distance between two *consecutive* positive peaks (or two *consecutive* negative peaks) of the waveform (Fig. 9.1) is known as the **wavelength** λ*. As time passes, the wave moves forward; Fig. 9.2 illustrates the positions of the wave at times t_1, t_2 and t_3. The number of cycles of the wave which pass a fixed point per second is known as the **frequency** ν† of the wave (8.1.3).

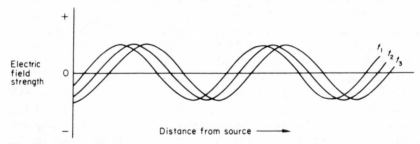

Fig. 9.2   The strength of the electric field plotted against distance from a source of electromagnetic radiation at times t_1, t_2 and t_3.

The third main characteristic of a wave is its **velocity** c, which is the distance travelled forward per second by a point on the wave; it is equal to the wavelength multiplied by the frequency. All electromagnetic waves, irrespective of their wavelengths or frequencies, travel at the *same* velocity in

*λ is the Greek small letter lambda.
†ν is the Greek small letter nu and should be distinguished from the English v in algebraic equations.

a given medium. In a vacuum this is about 3×10^8 metres per second; the velocity is very nearly the same in air.

The above relationship of the velocity of a wave to its wavelength and frequency is expressed in the equation:

$$\lambda v = c = 3 \times 10^8 \, \text{m s}^{-1}. \qquad \qquad \text{Eq. 9.1}$$

Note that $\lambda = c/v$, therefore a *high* (large) frequency corresponds to a *short* (small) wavelength and a *low* (small) frequency to a *long* (large) wavelength.

In the above, we considered the *electric* field in the electromagnetic wave. Instead, we could have considered the *magnetic* field, which varies in a similar manner. Both the electric and magnetic field strengths are *vector* quantities (1.3.2); consequently they have *direction* as well as magnitude. In an electromagnetic wave the directions of the electric and magnetic fields at a point are at right-angles to each other and they are both also at right-angles to the direction in which the wave is travelling.

9.1.3 The electromagnetic spectrum. There is a whole range of electro-magnetic radiations known as the **electromagnetic spectrum** (Table 9.1).

TABLE 9.1 (9.1.3) *The electromagnetic spectrum*

(1) Radiation	(2) Wavelength*	(3) Photon energy†
Radio, television and radar waves	3×10^4 m to 100 μm	$4 \cdot 1 \times 10^{-11}$ eV to $1 \cdot 2 \times 10^{-2}$ eV
Infra-red (heat rays)	100 μm to 700 nm	$1 \cdot 2 \times 10^{-2}$ eV to $1 \cdot 8$ eV
Visible light	700 nm to 400 nm	$1 \cdot 8$ eV to $3 \cdot 1$ eV
Ultra-violet	400 nm to 10 nm	$3 \cdot 1$ eV to 124 eV
X and gamma radiation	10 nm to 0·01 pm	124 eV to 124 MeV

*Units for wavelength are: m, metre; μm, micrometre (1 μm = 10^{-6} m); nm, nanometre (1 nm = 10^{-9} m), pm, picometre (1 pm = 10^{-12} m).

†Units for photon energy are: eV, electron-volt; MeV, mega-electron-volt (1 MeV = 10^6 eV).

Column 1: name of section of the electromagnetic spectrum,
Column 2: approximate range of wavelengths.
Column 3: corresponding range of photon energies (9.2.2).

They all have the characteristics described in section 9.1.2 and all travel at about $3 \times 10^8 \, \text{m s}^{-1}$ in a vacuum. They are produced in the various ways listed below, and differ in the magnitude of their wavelengths, in their properties and in the way in which they are detected.

The radiations in the various sections of the electromagnetic spectrum are produced in the following ways:

(i) **Radio, television and radar waves,** at the long wavelength end of the electromagnetic spectrum, are generated by high frequency alternating currents (i.e. decelerating electrons) flowing in the aerial of a radio or similar type of transmitter. They can be detected by placing a length of wire (i.e. a receiving aerial) in the path of the radiation; the waves induce an e.m.f. in the wire which is then amplified in the radio receiver.

(ii) **Infra-red rays,** or 'heat-rays', are radiated by the vibrating atoms or molecules in a *moderately* hot object (1.4.3) of which a good example is the hot element of a radiant electric fire. They can be sensed by the body as a feeling of warmth and detected either by photographic film or as an electric current generated when the radiation falls on a cell made from certain semiconductor materials similar to those used for solid-state rectifiers (11.3.2).

(iii) **Visible light** rays occupy only a narrow band of wavelengths and are so called because the human eye is sensitive to them. They can be produced in two ways: (a) they are emitted by the vibrating atoms or molecules in a *very* hot object and (b) they are emitted when orbital electrons jump between outer energy-levels in atoms or molecules. In addition to being sensed by the eye, they can be detected either by photographic film or by a photoelectric cell (13.4.2(c)). Visible light is dealt with in greater detail in section 9.3.

(iv) **Ultra-violet rays** have shorter wavelengths than those of visible light. They are generated mainly by the movement of orbital electrons between energy-levels in atoms. The process is similar to that for the production of visible light but involves higher energies and consequently shorter wavelengths (9.3.4). Ultra-violet rays cause pigmentary changes in the skin (e.g. sunburn) and can be detected either by photographic film or by certain types of photoelectric cell (13.4.2(c)) designed to be sensitive to this section of the electromagnetic spectrum.

(v) **X and gamma rays** are at the short wavelength end of the electromagnetic spectrum. X rays can be produced in two ways: (a) they are emitted when fast-moving electrons are decelerated and (b) they are emitted when orbital electrons jump between inner shells in atoms. Both processes can occur, for example, in the target of an X-ray tube; they are described in

Chapter 10. Gamma rays are emitted from the nuclei of atoms of some radioactive isotopes (2.3) and are described in Chapter 16. X and gamma rays can be detected in several ways using one of their various properties (10.1). Of particular importance is the detection of X and gamma rays by the ionization that they produce in air (15.1). Other methods of detection and measurement are described in section 15.5.

9.1.4 The inverse square law. Electromagnetic radiation travels in straight lines; this is known as **rectilinear propagation.** Consequently, if one takes a source of small physical size (i.e. a **point source**), the rays diverge in all directions from the point source in straight lines. Because the rays are spreading out, the intensity of the radiation (i.e. the energy flowing through unit area per unit time, 10.5.3) decreases with increasing distance from the source. The relationship between the intensity and the distance from the source is an **inverse square law** (compare section 3.1.2), provided that the reduction in intensity is due only to the geometrical divergence and not to any absorption or scattering of the rays (14.2.2) by the medium through which they are passing. This is seen from Fig. 9.3 which illustrates the divergence of rays from a point source at O. OP, OQ, OR and OS represent

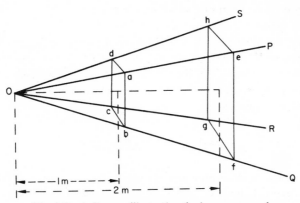

Fig. 9.3   A diagram illustrating the inverse square law.

the rays which pass through the corners a, b, c and d of unit area at 1 metre from O. At 2 metres from O, the same rays pass through the corners of the area represented by e, f, g and h. By the geometry of similar triangles, side ef is equal to twice side ab, and side fg is equal to twice side bc. Therefore area

115

efgh is *four times* the area abcd. As there is *no loss* of energy by absorption or scattering, *all* the energy passing through area abcd also passes through area efgh. Therefore the intensity (energy *per unit area* per unit time) at 2 metres is one-quarter of the intensity of 1 metre.

DEFINITION **The inverse square law states that the intensity of the radiation from a point source varies inversely as the square of the distance from the source, provided that there is no absorption or scattering by the medium.**

This law is represented by the equation:

$$\text{intensity} = \frac{k}{(\text{distance})^2},\qquad\qquad \text{Eq. 9.2a}$$

where k is a constant. This may also be expressed as:

$$\frac{\text{intensity at distance } d_1}{\text{intensity at distance } d_2} = \frac{1/d_1^2}{1/d_2^2} = \frac{d_2^2}{d_1^2}.\qquad\qquad \text{Eq. 9.2b}$$

For X and gamma radiation, the inverse square law is usually stated in terms of exposure rate and not intensity; the relationship between these two quantities is explained in section 10.5.3.

In practice, the inverse square law applies to X rays travelling through air if they are generated at voltages above about 50 kVp. At lower voltages, absorption and scattering by the air are not negligible; they cause the exposure rate to decrease with distance more rapidly than would be expected from the inverse square law.

9.2   THE QUANTUM THEORY OF RADIATION

9.2.1   The wave and corpuscular theories. So far we have described electromagnetic radiation in terms of a **wave theory**. However, certain properties of the radiation are better described in terms of a **corpuscular theory** in which the radiation is treated as a stream of particles or corpuscles. The two theories were first applied to the case of visible light. The wave theory gave a good account of the *transmission* of light but the corpuscular theory gave a better quantitative explanation of the way in which the light energy is *emitted* from a source and *absorbed* by a detector.

9.2.2 The quantum theory. The wave and corpuscular theories (9.2.1) were combined by Planck into one theory, the **quantum theory** of radiation. This states that energy is emitted or absorbed *only in small units* or 'packets' of energy known as quanta (singular: quantum). These quanta, which have no mass or electric charge, and which consist solely of energy, are also known as **photons.** The wave and corpuscular natures of the radiation are then related through the energy ϵ* of a quantum by the equation:

$$\epsilon = h\nu \qquad \text{Eq. 9.3}$$

where h = a constant known as Planck's Constant having a value of $6\cdot626 \times 10^{-34}$ joule seconds, and
ν = the frequency of the radiation.

Now, from Eq. 9.1 (9.1.2), $\lambda\nu = c$, therefore $\nu = c/\lambda$; hence, substituting in Eq. 9.3,

$$\epsilon = \frac{hc}{\lambda} \quad \text{or} \quad \epsilon \propto \frac{1}{\lambda}. \qquad \text{Eq. 9.4}$$

In words, this means that the quantum energy or photon energy ϵ is inversely proportional to the wavelength of the radiation.

Electromagnetic radiation can therefore be considered as a stream of quanta or photons, each photon having an energy equal to the product of Planck's Constant and the velocity divided by the wavelength of the radiation.

Column (3) of Table 9.1 (9.1.3) gives the photon energies corresponding to the wavelengths in Column (2). The photon energy corresponding to a given wavelength is calculated by substituting the values of Planck's Constant ($6\cdot626 \times 10^{-34}$ joule seconds), the velocity of the radiation ($3 \times 10^8 \, \text{m s}^{-1}$) and the wavelength *in metres* in Eq. 9.4, the value obtained for the energy being in joules, the S.I. unit of energy. The photon energies in Table 9.1, however, are given in **electron-volts** (eV), which is a unit of energy often more convenient to use in radiation work.

DEFINITION **The electron-volt is a unit of energy equal to the kinetic energy acquired by an electron when it is accelerated through a potential difference of 1 volt.**

One electron-volt is equal to $1\cdot602 \times 10^{-19}$ joules.

By substituting the values of Planck's Constant and the velocity in Eq. 9.4 and by changing the units for photon energy and wavelength, one obtains the important relationship (10.4.3):

*ϵ is the Greek small letter epsilon.

$$\text{photon energy in keV} = \frac{1\cdot24}{\text{wavelength in nanometres}}, \qquad \text{Eq. 9.5}$$

where $1\,\text{keV} = 10^3\,\text{eV}$ and 1 nanometre $= 10^{-9}$ m.

9.3 VISIBLE LIGHT AND FLUORESCENCE

9.3.1 The production of light. In section 9.1.3 we saw that visible light is electromagnetic radiation having wavelengths in the small range over which the eye is sensitive to such radiation. This range extends from about 700 nanometres to 400 nanometres ($1\,\text{nm} = 10^{-9}$ m).

Visible light can be produced in two ways. First, when a solid body is heated, the atoms and molecules vibrate and the vibration results in the emission of photons (1.4.3). If the temperature of the body is sufficiently high, some of the radiation emitted has wavelengths in the visible region of the spectrum (9.3.3). Second, photons are emitted when electrons jump between outer energy-levels in atoms or molecules following ionization or excitation. Again, some of these photons have wavelengths in the visible region of the spectrum (9.3.4).

9.3.2 Types of spectrum. The **spectrum** of a beam of radiation is a description of the range of wavelengths present in the beam and of their relative intensities. For visible light, there are two types of spectrum:

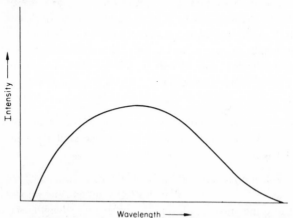

Fig. 9.4   An example of a continuous spectrum in which the intensity of the radiation is distributed in a continuous fashion over all the wavelengths present.

(a) the **continuous spectrum** in which the intensity of the light is distributed in a *continuous* fashion over the range of wavelengths present (Fig. 9.4),

(b) the **line or characteristic spectrum** in which the light photons occur at *a few wavelengths only* (Fig. 9.5).

Visible light can be produced with a continuous spectrum (9.3.3) or with a line spectrum (9.3.4), or with both types of spectrum occurring together. This differs from the output of an X-ray tube, in which the line spectrum of the X rays cannot occur alone and, if present, is *superimposed* on the continuous spectrum (10.4.1).

9.3.3 The continuous spectrum of light results from the vibrations of the atoms and molecules of a solid body when it is heated to a sufficiently high temperature (1.4.3). The photons radiated by the vibrating atoms and molecules have a range of wavelengths; the shortest wavelength present (i.e. the highest photon energy) depends on the temperature to which the body is raised. If the temperature is moderate, say $400°C$, the shortest wavelength radiated is in the infra-red region of the electromagnetic spectrum and the body radiates heat rays but no visible light. If the temperature is raised to say $700°C$, the shortest wavelength present is now in the visible region of the spectrum and the body glows red, i.e. it radiates visible light as well as infra-red rays. If the temperature is raised further, the atoms and molecules vibrate more vigorously and the wavelengths of some of the photons radiated

TABLE 9.2 (9.3.3) *The approximate relationship between the wavelength of the light and the colour seen by the eye*

Wavelength in nanometres*	Colour
700	Red
600	Orange
550	Yellow
500	Green
450	Blue
425	Indigo
400	Violet

* 1 nanometre = 10^{-9} metre.

A useful mnemonic for remembering the colours in the spectrum of visible light, in order of decreasing wavelength, is **Richard Of York Gave Battle In Vain: Red Orange Yellow Green Blue Indigo Violet.**

are shorter still and the body glows yellow. If the temperature is raised to say 1 500° C, the body appears white and is described as 'white hot'.

The change in colour of the body arises because photons of progressively shorter wavelengths (i.e. higher photon energies) appear to the eye as red, orange, yellow, green, blue, indigo and violet. The relationship between the wavelength of the radiation and the colour seen by the eye is given in Table 9.2 When the body is at a high temperature such as 1 500° C, photons of all wavelengths in the visible region are present (i.e. all the colours are present) and the radiation appears to the eye as **white light**.

From the above, it is seen that the minimum wavelength in the continuous spectrum of the light decreases as the temperature of the body increases. This is analogous to the case for X rays where the minimum wavelength in the continuous spectrum decreases with increase in the peak value of the voltage across the tube (10.4.3).

9.3.4 The line or characteristic spectrum. When an atom is excited or ionized (2.4) by giving energy to an electron in an *outer* shell or energy-level, the vacancy created is subsequently filled by the **transition** or jump of an electron to that shell or energy-level (from one farther out) with the emission of a photon. As *outer* energy-levels of the atoms are involved, the energy of the photon (which is equal to the difference between the binding energies of

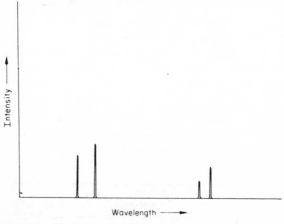

Fig. 9.5   An example of a line or characteristic spectrum in which the radiation occurs at a few wavelengths only.

the two energy-levels involved) is such that it is in the visible or ultra-violet regions of the spectrum. This process gives rise to the **line or characteristic spectrum**. The energy of the photon produced (and hence the wavelength of the radiation), being dependent on the binding energies of the levels involved in the transition, is consequently *characteristic* of the element of which the atom is a part. The process is similar to that which gives rise to the characteristic spectrum of X rays but in the latter case the *inner* shells of atoms are involved and consequently the photons radiated have higher energies which bring them into the X-ray region of the spectrum (10.4.4).

A common example of a characteristic spectrum is the yellow light radiated by sodium-vapour street lamps. Atoms of sodium are excited by an electrical discharge in the lamp and yellow light characteristic of the sodium atom is emitted during the subsequent transitions of the electrons.

9.3.5 Fluorescence is the property possessed by certain crystalline substances of emitting characteristic radiation in the visible or ultra-violet regions of the spectrum after absorbing electromagnetic radiation of shorter wavelength. The short-wavelength radiation produces excitation in the substances and the subsequent transitions of electrons are accompanied by the emission of radiation that is characteristic of the crystals.

Fluorescence has important applications in radiology. The simple viewing screen used in the traditional technique for fluoroscopy (before the days of image intensifiers) consisted of a layer of **zinc-cadmium sulphide** crystals supported on a white card and placed behind a sheet of lead glass. The X rays caused the crystals to fluoresce, emitting a yellow-green light to which the eye is very sensitive. The lead glass is relatively transparent to this visible light but it strongly attenuates the X rays thus providing adequate shielding for the radiologist viewing the screen.

Although traditional fluoroscopy was very valuable because it displayed movement in the patient, it had the disadvantage that for X-ray intensities replaced in fluoroscopy by systems called **X-ray image-intensifier television** tated long periods of dark-adaptation for the radiologist's vision; even then he was unable to see the whole of the information displayed on the screen because of the poor efficiency of his vision at such low brightness levels.

In the past twenty years the simple fluorescent screen has been gradually replaced in fluoroscopy by systems called **X ray image-intensifier television systems**. In these, the X rays are again converted into light by means of a fluorescent screen, but the latter is enclosed in a device called an **X-ray image**

121

intensifier. This, by electronic means, produces an image at brightness levels of about 5000 times that of the simple fluorescent screen. Moreover, this image is viewed via a small television camera connected to one or more **television monitors** (receivers) so that it can be clearly seen at a number of different locations in the X-ray room or even elsewhere, and can be electronically recorded for future examination.

Another important application of fluorescence is the use of a pair of fluorescent **intensifying screens** in the cassette that holds the film used in radiography (10.1 (ii)). Each screen consists of a layer of **calcium tungstate** or other suitable fluorescent material (such as the so-called **rare-earth** compounds) that fluoresce under the action of the X rays with the emission of ultra-violet, violet, blue or green light. About 95% of the image formed on the photographic film is due to the fluorescent light and only about 5% to the direct action of the X rays on the photographic emulsion, the precise proportions depending on the particular screen/film combination being used.

10 X Rays

10.1 THE PROPERTIES OF X RAYS

X rays were discovered by Röntgen in 1895 when he was investigating the conduction of electricity through gases at low pressure in glass tubes. He noticed that the positive electrodes in the tubes gave off invisible rays that caused fluorescent screens to glow and that fogged photographic plates. The rays were very penetrating; they passed through black paper and even thicker objects. They were not deflected in a magnetic field; Röntgen therefore concluded that they were not streams of electrically charged particles which would have been deflected like a conductor carrying a current (6.2.2). As their nature was unknown, he called them 'X rays'. Later, in 1912, they were shown to be electromagnetic radiation (Chapter 9) of very short wavelength.

The properties of X rays may be grouped under the following headings:

(i) **Fluorescence** (9.3.5). X rays produce fluorescence in materials such as calcium tungstate, zinc cadmium sulphide and caesium iodide. This effect produces the visible pattern seen on the simple screen in X-ray fluoroscopy ('screening') and is utilized in intensifying screens (see (ii) below).

(ii) **Photographic effect.** X rays produce a latent image on photographic film which can be developed to give a visible image as in an ordinary photographic negative. This direct effect is utilized in film-badge dosimetry for radiation protection (17.5.2) and to a small extent in radiography. A radiographic film is most often used in a cassette containing a pair of fluorescent intensifying screens (9.3.5). Only about 5% of the image is formed by the *direct* action of the X rays; the remaining 95% is due to the visible and ultra-violet light emitted by the intensifying screens and to which the film is more sensitive. The precise proportions depend on the particular screen/film combination being used.

(iii) Penetration. X rays penetrate substances that are opaque to visible light. They are gradually absorbed the farther they pass through an object; the amount of the absorption depends on the atomic number and the density of the object and on the energy of the X rays (Chapter 14). An understanding of the way in which X rays are absorbed has several important applications: diagnostic radiology is based on differences of absorption in body structures, radiotherapy requires the calculation of the doses of radiation absorbed by parts of the body, and radiological protection involves the design of shielding to absorb radiation.

(iv) Ionization and excitation (2.4). X rays produce ionization and excitation of the atoms and molecules of the substances through which they pass (14.1). These processes are important in all interactions of X rays with matter and form the basis of the properties (i) to (vi) listed here. The ionization of air by X rays passing through it can be demonstrated by irradiating the air surrounding a gold-leaf electroscope (3.2.1) which has been electrically charged. Air is normally a good insulator; however, it becomes a conductor of electricity when ionized. The leaves of the electroscope then fall together as the electric charge leaks away through the ionized air. The standard method of measuring quantity of X radiation or **exposure** is based on the ionization of air. The unit of exposure, the roentgen, is defined in section 15.2.1.

(v) Chemical changes. X rays produce chemical changes in substances through which they pass (14.1). One important change is the oxidation of ferrous sulphate ($FeSO_4$) in solution to ferric sulphate. The amount of ferric sulphate ($Fe_2 (SO_4)_3$) produced can be used as a measure of the quantity of X radiation absorbed and is the basis of the chemical system of dosimetry named after Fricke (15.5.7).

(vi) Biological effects. X rays produce biological effects in living organisms, either by *direct* action on the cells or *indirectly* as a result of chemical changes near the cells (14.1.2). The cells can be either damaged or killed. As a result, the organism itself can be injured or killed, or in the reproductive cells the genes can undergo mutations resulting in inherited changes in subsequent generations. Biological effects have to be considered in various contexts, for example, the need to protect individuals from overexposure (17.2), the planned killing of malignant tumour cells in radiotherapy, and the sterilization of hospital supplies, such as syringes and dressings, by large doses of radiation.

10.2 THE PRODUCTION OF X RAYS

10.2.1 Energy loss by electrons. X rays are produced when electrons give up energy by either one of two processes: (i) the deceleration of a fast-moving electron resulting in the conversion of some of its kinetic energy into X-ray energy, and (ii) the movement of an electron between two inner shells in an atom with the difference between the binding energies of the two shells being radiated as an X-ray photon. Both these processes can occur in the target of an X-ray tube (10.3).

10.2.2 The principles of operation of an X-ray tube. In a modern X-ray tube, which is sometimes called a Coolidge tube after the inventor, electrons are released from a heated filament by thermionic emission (11.1.1). The electrons are then accelerated across the tube by a high voltage applied between the filament and the anode. When the electrons reach the anode, they are travelling at a high velocity; therefore they have high kinetic energy and this is converted into X rays and heat as the electrons interact with the atoms of the anode (10.3).

The main features of an X-ray tube are shown diagrammatically in Fig. 10.1; further constructional details and the electrical principles of operation are described in Chapter 11.

(i) **The filament,** which is usually a spiral of tungsten wire, is heated by passing an electric current through it from a **low-voltage supply.** Electrons are then released from the filament by **thermionic emission** (11.1.1). Tungsten is used because it produces appreciable thermionic emission at temperatures well below its melting point.

(ii) **A high-voltage supply** is connected between the filament which acts as the cathode and the **target** which is part of the anode of the tube. The targets of tubes used in medicine are usually made of tungsten; it has a high melting point, adequate thermal conductivity, and a high atomic number (74) which increases the efficiency of X-ray production (10.6.4).

(iii) **Kinetic energy** is gained by the (negatively charged) electrons released from the filament as they are accelerated to high velocity by the positive voltage applied to the target. This energy is then given up by the electrons in interactions with the target (10.3).

(iv) A high vacuum exists in the tube so that there is no gas present to produce electrons by ionization as was the case in Röntgen's original experiments (10.1). In these, electrons released by ionization of the gas molecules in the tube produced the X rays at the positive electrode.

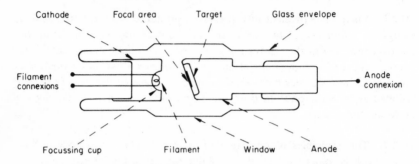

Fig. 10.1   A schematic diagram of an X-ray tube with a stationary anode.

(v) A shield or focussing cup mounted near the filament forms part of the cathode assembly. This protects adjacent parts of the tube wall from damage by electron bombardment and is so shaped that it produces an electric field that focusses the electrons on to a small area of the target known as the **focus** or **focal area**. It is from the focal area, therefore, that the X rays emerge.

10.3   INTERACTIONS OF ELECTRONS WITH THE TARGET

The kinetic energy of the electrons when they reach the target of an X-ray tube is proportional to the value of the high voltage which has accelerated them across the tube. This energy is converted into heat and X rays as the electrons interact or 'collide' with atoms in a thin surface layer of the target. An average of less than 1% of the energy brought by the electrons to the target of a diagnostic tube is converted into X-ray energy. In other words, the efficiency (5.2.2) of X-ray production is less than 1%. This is because processes (i) and (ii) below predominate. At much higher voltages the efficiency of X-ray production is far greater; for example when X rays are produced at 4 MeV in a linear accelerator, the efficiency is about 40%.

Four types of interaction are possible when an electron arrives at the target (Fig. 10.2):

(i) **Excitation involving an electron in an outer shell** of an atom in the target. The incident electron coming from the filament transfers a small amount of energy (only a few electron volts) to an electron in an *outer* shell of an atom in the target and displaces it to an energy level farther out. The

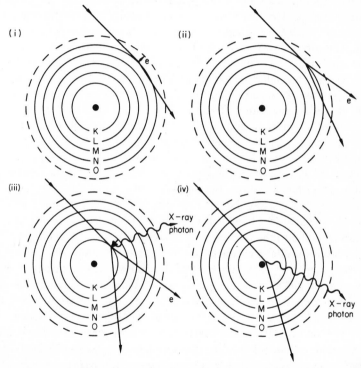

Fig. 10.2 The possible interactions of the electrons from the cathode with the atoms in the target of an X-ray tube: (i) excitation, (ii) ionization, (iii) ionization followed by the emission of a characteristic X-ray photon, and (iv) Bremsstrahlung production.

process is that of excitation (2.4). The electron returns to the vacancy in the shell and the energy released in this transition appears as heat in the target.

(ii) **Ionization by the removal of an electron from an outer shell** of an atom in the target. The incident electron transfers sufficient energy to ionize an atom of the target by the removal of an electron from an *outer* shell (2.4).

The displaced electron, known as a **secondary electron,** may produce further ionization or excitation in other atoms of the target. Again only a small amount of energy is given up by the incident electron and it ultimately appears as heat.

(iii) **Ionization by the removal of an electron from an inner shell** of an atom in the target and the subsequent emission of a characteristic X-ray photon. The incident electron transfers sufficient energy to remove an electron from an *inner* shell of an atom in the target. To do this, the incident electron must have energy *equal to or greater than* the binding energy for that shell. The difference between the binding energy and the amount of energy transferred from the incident electron is carried away by the displaced electron (the **secondary electron**) as kinetic energy. This kinetic energy is given up by the secondary electron as it produces ionization and excitation of other atoms in the target. The vacancy in the inner shell is filled by an electron moving inwards from another shell in the atom. When this **transition** or jump occurs, it is accompanied by the emission of an X-ray photon of energy equal to the difference between the binding energies of the two shells involved in the transition. This photon is known as a **characteristic X-ray photon.** It is *characteristic* of the element of which the target is made because its energy is related to the binding energies of shells in an atom of that particular element (the binding energies differ from element to element). This process gives rise to the **characteristic X-ray spectrum** (10.4.4).

(iv) **Bremsstrahlung production.** The incident electron passes close to the *nucleus* of an atom in the target. The electron is negatively charged, and the attraction of the positive electric charge on the nucleus makes it decelerate (1.3.2). Consequently electromagnetic radiation must be emitted (9.1.1) and the electron loses energy in the form of an X-ray photon. The energy of the X-ray photon depends on the degree to which the electron is decelerated by the attraction of the nucleus; the photon energy can take any value from zero to a maximum. The latter occurs when the electron passes very close to the nucleus and the deceleration is so great that the electron comes to rest. *All* its kinetic energy is thus converted in the one interaction into the energy of a *single* X-ray photon (10.4.3).

The X radiation produced by the deceleration of electrons is known as **Bremsstrahlung** (i.e. German for 'braking radiation') and it gives rise to the **continuous spectrum of X rays** (10.4.2).

10.4 SPECTRA OF X RAYS

10.4.1 Types of spectrum. For visible light there are two types of spectrum (9.3.2). Similarly for X rays there are:

(a) the **continuous spectrum**, in which the intensity of the X rays is distributed in a *continuous* fashion over the range of wavelengths present, and

(b) the **line or characteristic spectrum**, in which the X rays occur at *a few wavelengths only*.

When there is a characteristic spectrum of X rays present in the output of an X-ray tube, it appears as peaks *superimposed* on the smooth curve of the continuous spectrum (Fig. 10.3); it cannot be produced separately as in the case of light (9.3.2). A characteristic spectrum of X rays can, however, be produced separately by other methods.

10.4.2 The continuous spectrum of X rays is also known as **general** or **white radiation** by analogy with white light (9.3.3). The continuous spectrum is the result of Bremsstrahlung production (10.3(iv)) and it has the following main features (spectrum (i) in Fig. 10.3a, 10.4.1):

(1) A definite short wavelength limit λ_{min} to the spectrum (10.4.3).

(2) All wavelengths greater than the short wavelength limit are present in the radiation until the long wavelength limit λ_{max} is reached, i.e. the spectrum is continuous.

(3) The long wavelength limit is not as clearly defined as the short wavelength limit. X rays of all wavelengths longer than the minimum are generated in the target of an X-ray tube, but those of very long wavelength do not emerge from the tube assembly because of attenuation: (a) in the target itself, (b) in the materials of the tube and of the window in the tube housing, (c) in the cooling oil, and (d) in any added filters (14.5).

(4) A peak in the intensity occurs at a wavelength two to three times the minimum wavelength for the usual amounts of filtration.

In Fig. 10.3a intensity is plotted against *wavelength*. If intensity is plotted against *photon energy* instead (Fig. 10.3b), the graph will appear 'left-to-right' because short wavelength corresponds to high energy, etc., with the minimum wavelength λ_{min} corresponding to the maximum photon energy ϵ_{max}. (Strictly, the wavelength is inversely proportional to the photon energy, 9.2.2.)

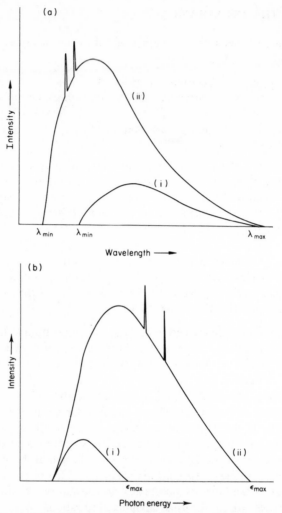

Fig. 10.3   X-ray spectra: (i) the continuous spectrum only, and (ii) the continuous spectrum with the characteristic line spectrum superimposed. (a) intensity plotted against wavelength, (b) intensity plotted against photon energy.

10.4.3 The short wavelength limit. X-ray photons of the **minimum wavelength (maximum photon energy)** in the spectrum are produced only when both of the following occur to *one* electron:

(a) it is accelerated across the X-ray tube to the target by the peak value of the applied voltage,

(b) it passes so close to the nucleus of an atom in the target that it is decelerated to rest with all its kinetic energy being converted into a *single* X-ray photon.

The energy of this photon (the maximum photon energy in the spectrum) is consequently proportional to the peak value of the applied voltage; its wavelength, the *minimum wavelength* in the spectrum, *is inversely proportional to the peak voltage*. Note that the minimum wavelength does *not* depend on the material of which the target is made.

The minimum wavelength in a spectrum is calculated as follows:

let λ_{min} = the minimum wavelength,

ϵ_{max} = the (maximum) photon energy corresponding to λ_{min},

ν = the frequency corresponding to λ_{min},

c = the velocity of electromagnetic radiation in vacuum (about 3×10^8 m s^{-1}),

h = Planck's Constant ($6 \cdot 626 \times 10^{-34}$ joule . second) (9.2.2),

V_p = the peak value of the applied voltage,

e = the charge on an electron ($1 \cdot 602 \times 10^{-19}$ coulomb).

The energy of an electron arriving at the target after being accelerated by the peak voltage = eV_p (Eq. 5.1, 5.1.1). But this is all converted into the energy of the photon of minimum wavelength,

i.e.

$$eV_p = \epsilon_{max}.$$

Now, applying the quantum theory (Eq. 9.4, 9.2.2),

$$\epsilon_{max} = \frac{hc}{\lambda_{min}},$$

therefore

$$eV_p = \frac{hc}{\lambda_{min}} \quad \text{or} \quad \lambda_{min} = \frac{hc}{eV_p}.$$

h, c and e are constants and if their values are substituted in the equation, we get

$$\lambda_{min} = \frac{1 \cdot 24}{\text{peak voltage in kilovolts}} \text{ nanometres}, \qquad \text{Eq. 10.1}$$

where 1 nanometre = 10^{-9} metre.

This important relationship is sometimes known as the **Duane-Hunt Law.**

In some former textbooks, wavelength is expressed in Ångström units ($1 \text{ Å} = 10^{-10}$ metre) and Eq. 10.1 becomes

$$\lambda_{min} = \frac{12 \cdot 4}{\text{peak voltage in kilovolts}} \text{ Å}.$$

10.4.4 The line or characteristic spectrum is *superimposed* on the continuous spectrum of an X-ray beam (spectra (ii) in Figs 10.3a and b, 10.4.1) if the applied voltage across the tube is *equal to or greater than* the critical voltage for the production of characteristic radiation in the particular material of which the target is made. Electrons which have been accelerated across the tube by the critical voltage arrive at the target with energy equal to the binding energy of an electron in the K-shell of an atom in the target; consequently atoms are ionized by the removal of an electron from the K-shell. The vacancy is then filled by the transition of an electron from another shell, usually the L- or M-shell, to the K-shell. The transition is accompanied by the emission of an X-ray photon of energy equal to the difference between the binding energies of the two shells involved.

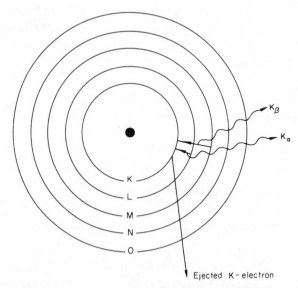

Fig. 10.4 An illustration of the transitions which produce the K_α and K_β lines in the characteristic spectrum.

In the target, many atoms are ionized by the removal of an electron from the K-shell; in some the vacancy is filled by a transition of an electron from the L-shell of the atom, in others by a transition from the M-shell, and so on. Therefore a *series* or *group* of characteristic lines (the **K-series**) appears in the spectrum arising from the various transitions that can take place.

The transitions giving rise to the K-series are illustrated in Fig. 10.4 for the atom of tungsten (which is the usual target material in medical X-ray tubes). The X-ray photon emitted by an electron moving from the L-shell to fill the vacancy in the K-shell is known as K_α **radiation***; the transition of an electron from the M-shell to the K-shell gives rise to K_β **radiation***. Similarly, the **L-series** of lines is emitted if electrons are ejected from the L-shells of the atoms. The subsequent transition of an electron from the M-shell to the L-shell gives rise to L_α **radiation**, etc.

The photons of the K-series are usually the only characteristic radiations with sufficient energy to emerge from a medical X-ray tube; of these, the K_α and K_β are the most prominent. (The photons of the L-series have smaller energies and are almost completely absorbed before emerging from the window of the tube.) The energies of the characteristic X-ray photons, being dependent on the binding energies, are greater the higher the atomic number of the target material, because the binding energies increase with increasing atomic number. For tungsten, for example, the photon energies of the K_α and K_β radiations are respectively about 59 and 69 keV which correspond to wavelengths of 0·021 and 0·018 nanometre.

As the critical voltage for production of K characteristic radiation depends on the binding energy of the electrons in the K-shell of atoms in the target, it increases with increasing atomic number of the target. *For tungsten, the critical voltage is about 70 kV.*

10.5   THE QUALITY AND INTENSITY OF X RAYS

10.5.1   The quality of an X-ray beam. It is necessary to be able to describe the *quality* of a beam of radiation (i.e. to describe how penetrating it is) as well as to state the *amount* or *quantity* involved.

With **homogeneous (monochromatic** or **monoenergetic) radiation**, where there is only a single wavelength present, i.e. where all the photons have the same energy, the **quality** is completely described by stating the wavelength or

* a and β are the Greek small letters alpha and beta respectively.

the photon energy. The beam from an X-ray tube, however, is **heterogeneous**, i.e. many wavelengths or photon energies are present. To describe *fully* the quality of such a beam, the spectrum of the radiation must be given as in Fig. 10.3 (10.4.1), i.e. the relative intensities of radiation of each wavelength or photon energy must be stated. In radiology, however, it is not often necessary for the description to be as detailed as this and the **quality** is usually specified by stating either the **half-value layer** or the **effective photon energy** of the radiation (10.5.2) together with the value of the **applied voltage** and the **filtration**.

10.5.2   Half-value layer; effective photon energy. For some applications, the quality of a beam of X or gamma radiation can be adequately described in terms of its **half-value layer** (h.v.l.), sometimes called **half-value thickness** (h.v.t.). (For the measurement of h.v.l., see section 15.3.)

DEFINITION   **The half-value layer or half-value thickness of a beam of radiation is that thickness of a stated material which reduces the exposure rate (10.5.3) of the beam to one half.**

The material specified is chosen so that the half-value layer is a convenient thickness for measurement, say a few millimetres. Aluminium is commonly used for X radiation generated by voltages up to 120 kVp and copper for higher voltages.

The quality of a heterogeneous beam can also be described in terms of its **effective photon energy** (or **effective wavelength**).

DEFINITION   **The effective photon energy (or effective wavelength) of a heterogeneous beam of radiation is the photon energy (or wavelength) of that homogeneous beam which has the same half-value layer as the heterogeneous beam.**

The half-value layer or the effective photon energy does not give an unambiguous description of the quality of a beam of radiation; two X-ray beams with different spectra (and therefore of *different qualities*) can have the same half-value layer but their exposure rates would be reduced by unequal amounts by *other* thicknesses of the same material (Fig. 10.5). Consequently, in order to describe more precisely the quality of a heterogeneous beam, the peak value of the applied voltage *and the filtration* (14.5) are often stated in addition to the half-value layer or the effective photon energy.

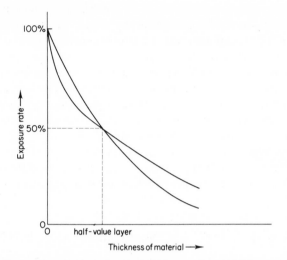

Fig. 10.5   Exposure rate plotted against thickness of material for two X-ray beams of different qualities but with the same half-value layer in the material.

10.5.3   Intensity and exposure rate. We can express the *quantity* of radiation flowing per unit time in terms of the **intensity** of the beam of X rays.

DEFINITION   **Intensity is the amount of energy flowing per unit time through unit area of a plane normal to (i.e. 'at right-angles to') the direction of propagation.**

In the spectra shown in Fig. 10.3a (10.4.1), the intensity of X rays of each wavelength is plotted against wavelength. The *total* intensity in the beam is then the sum of the intensities at all the wavelengths present and is represented in Fig. 10.3a by the area under the curve, i.e. between the curve and the wavelength axis.

The intensity of a beam is a measure of the energy (1.3.5) *flowing through* unit area per unit time ($J\,m^{-2}\,s^{-1}$) whereas in radiology we are interested in the *effects* produced by the radiation; these result from the energy *absorbed* by the medium (14.1). Quantity of radiation could be expressed in terms of *any* of the effects produced; *one* effect which is particularly convenient is the ionization of air by X radiation (15.1). Hence we have **exposure** (formerly called exposure dose); this is a measure of quantity of X radiation based on

135

the ability of the radiation to ionize the air through which it passes (15.2.1) and is proportional to the energy absorbed by the air. The unit of exposure is the **roentgen** (15.2.1); **exposure rate**, i.e. exposure per unit time, is measured in roentgens/second, roentgens/minute, etc.

The exposure rate is related to the intensity of the beam by the real absorption coefficient for the radiation in air (14.3.8). *The exposure rate is consequently proportional to the intensity for a given quality of radiation.*

The word 'exposure' is sometimes used in another sense; it can refer to the product of the tube current and the time interval for which the X-ray tube is energized when taking a radiograph or giving an exposure in radiotherapy, i.e. to mA × seconds (13.5.2). It is important to recognize these different uses of the word when making calculations such as those in examples A and B in section 10.6.5.

10.6   THE FACTORS INFLUENCING QUALITY AND INTENSITY

10.6.1   The voltage applied across the X-ray tube. The value of the applied voltage affects both the quality and the intensity of the X rays produced (Fig. 10.6). As the applied voltage is increased, the spectrum extends to higher pho-

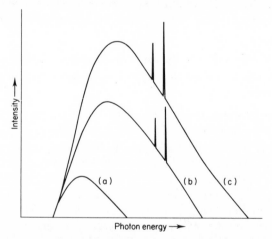

Fig. 10.6   The influence of the applied voltage on the spectrum of X rays. (a) A small applied voltage (less than the critical voltage for the production of the characteristic line spectrum); (b) a larger applied voltage (greater than the critical voltage for the production of the characteristic line spectrum); and (c) a still larger applied voltage.

ton energies resulting in increases in the half-value layer and the effective photon energy of the radiation. (The maximum photon energy is proportional to the peak value of the applied voltage, 10.4.3). The intensities at all photon energies present increase as the applied voltage is increased; this results in the *total* intensity, given by the area under the curve, being *approximately* proportional to the applied voltage squared.

Note (i) If the applied voltage exceeds the critical voltage (10.4.4), the characteristic line spectrum will appear; however the value of the applied voltage does *not* affect the *photon energies* at which the characteristic lines occur. These depend on the binding energies of the shells in the atoms of the target material and hence only on the atomic number of the target material.

(ii) The quality of an X-ray beam depends also on the type of rectification used in the high voltage supply (12.4.2). If the instantaneous value of the applied voltage is constant, as in a set with a constant potential generator (12.5.2), the spectrum of the X-ray beam during every small interval of time will be the same. If, however, the instantaneous value of the applied voltage is not constant, e.g. as in a set with self-rectification (12.2.1) or with pulsating voltage (12.3.1), the spectrum changes from moment to moment as the voltage across the tube changes. Consequently, the shape of the average or overall spectrum in the latter case is different from the shape of the spectrum pro-

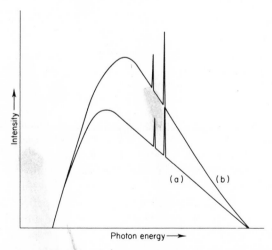

Fig. 10.7 The influence of the type of rectification on the spectrum of X rays: (a) pulsating voltage and (b) constant potential.

duced at constant potential because it includes a greater proportion of low energy radiation (Fig. 10.7).

10.6.2 The tube current. The value of the tube current, i.e. the flow of electrons from the filament to the target, affects the intensity but *not* the quality of the X rays produced (Fig. 10.8). The intensities at all the photon energies present increase in proportion to the tube current. There is therefore no change in the shape of the spectrum or in the maximum photon energy but the total intensity is proportional to the average tube current.

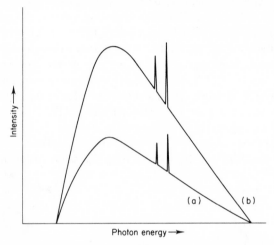

Fig. 10.8 The influence of the tube current on the spectrum of X rays. The tube current producing spectrum (b) is twice that producing (a).

10.6.3 Filters are employed to alter the quality of a beam of X rays but they also reduce the total intensity of the beam. A simple filter consists of a thin sheet of material in which the attenuation increases rapidly with decrease in photon energy (at the energies commonly used in radiology) so that there is much greater attenuation of the low photon energy (long wavelength) radiation than of the high photon energy (short wavelength) radiation in the beam (14.5). The shape of the spectrum is therefore altered (Fig. 10.9).

The filtered beam is described as *harder* (i.e. more penetrating) than the unfiltered beam because it contains a higher proportion of high photon energy radiation which is the more penetrating.

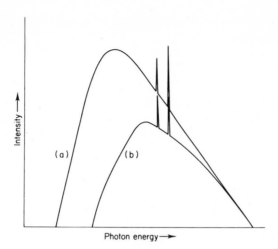

Fig. 10.9 The influence of filtration on the spectrum of X rays: (a) before, and (b) after filtration.

10.6.4   The atomic number of the target material. The atomic number of the target material affects the intensity of the continuous spectrum of the X rays produced (Fig. 10.10). The intensities at all the photon energies increase with increase in atomic number so that the total intensity is approximately proportional to atomic number, i.e. X-ray production is more efficient the higher the atomic number of the target material. This is because the amount of Bremsstrahlung produced (10.3(iv)) increases with increasing atomic number of the target material. Note that the maximum photon energy (minimum wavelength) in the spectrum *does not change* with atomic number and that there is no change in the quality of the Bremsstrahlung.

When selecting the material for the target of an X-ray tube, the melting point has to be considered in addition to the atomic number. In practice, tungsten (atomic number 74, melting point 3370° C) is almost always used in medical tubes although gold (atomic number 79, melting point 1063° C) is used in some megavoltage machines. In the latter, less heat is generated in the target because X-ray production is more efficient at these very high voltages (10.3).

The atomic number of the target material also affects the critical voltage and the *photon energies* of the lines in the characteristic spectrum (10.4.4). An increase in atomic number of the target material results in a higher critical

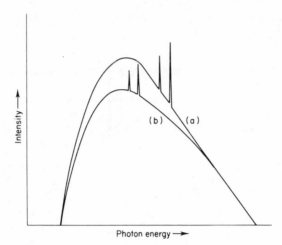

Fig. 10.10 The influence of the atomic number of the target material on the spectrum of X rays. Spectrum (a) is produced by a target material of higher atomic number than that which produces (b).

voltage and in higher photon energies (shorter wavelengths) for the lines in the characteristic spectrum (Fig. 10.10).

10.6.5 Summary and examples. The **quality** of a beam of X rays depends on:

      (a) the peak value of the applied voltage,

      (b) the type of rectification,

      (c) the filtration,

      (d) the presence of characteristic radiation.

The **intensity** or the **exposure rate** is proportional to:

      (a) the applied voltage squared (approximately),

      (b) the tube current,

      (c) the atomic number of the target material.

In addition, the total intensity is reduced by filtration.

In symbols,

$$\mathscr{I} \propto V^2 I Z,$$ Eq. 10.2

where

      $\mathscr{I}$ = exposure rate (which is proportional to intensity),

      V = applied voltage,

      I = tube current,

      Z = atomic number of target material.

If the exposure rates at different distances d from the focus of the X-ray tube are to be compared, the inverse-square law (9.1.4) can be combined with Eq. 10.2 to give

$$\mathcal{J} \propto \frac{V^2 IZ}{d^2}.$$ Eq. 10.3

To obtain the corresponding equation for exposure (quantity of X radiation), the time interval t must be introduced into Eq. 10.3 to give

$$\text{exposure} \propto \mathcal{J} t \propto \frac{V^2 IZt}{d^2}$$ Eq. 10.4

Note that the target material for a medical X-ray tube is almost always the same (tungsten) and so the atomic number could be deleted from Eqs. 10.3 and 10.4.

Example A. Consider an X-ray tube which produced an exposure of 20 roentgens in 60 seconds at a point 0·5 metre from the focus of the tube when working with an applied voltage of 100 kVp and a tube current of 15 mA. How long would it take to produce an exposure of 10 roentgens at a point 0·4 metre from the focus when the tube is working at 80 kVp and 5 mA?

Substitute in Eq. 10.4 for the first set of conditions putting $V = V_1$, etc.

$$\text{first exposure} \propto \frac{V_1^2 I_1 Z_1 t_1}{d_1^2}.$$ (1)

Similarly for the second set of conditions

$$\text{second exposure} \propto \frac{V_2^2 I_2 Z_2 t_2}{d_2^2}.$$ (2)

Dividing (2) by (1) gives

$$\frac{\text{second exposure}}{\text{first exposure}} = \frac{V_2^2 I_2 Z_2 t_2 d_1^2}{V_1^2 I_1 Z_1 t_1 d_2^2}$$ (3)

Note that as (1) and (2) are proportionalities, when (2) is divided by (1) the proportional sign is replaced by an equals sign in the resulting equation (3).

Substituting numerical values in (3) gives

$$\frac{10}{20} = \frac{(80)^2}{(100)^2} \frac{5Z_2 \, t_2}{15Z_1} \frac{(0·5)^2}{60(0·4)^2}.$$ (4)

But $Z_1 = Z_2$ (because the target material is unchanged) and they cancel out in the equation.

Rearranging (4) to make t_2 the subject of the equation gives

$$t_2 = \frac{10(100)^2 \; 15(0{\cdot}4)^2 \; 60}{20(80)^2 \; 5(0{\cdot}5)^2} = 90 \text{ seconds.}$$

The time taken to produce an exposure of 10 roentgens under the new set of operating conditions is therefore 90 seconds.

Example B. As an example of the use of the word 'exposure' to mean the product of tube current and time (10.5.3), consider a radiograph taken at 2 metres focus–film distance and requiring an 'exposure' of 15 mA seconds. What 'exposure' would give the same photographic density if the focus–film distance was reduced to 1 metre, all other factors remaining the same?

In order to produce the same photographic density, the quantity of X radiation reaching the film must be the same under both sets of conditions.

From Eq. 10.4, exposure (first meaning) = quantity of X radiation $\propto (V^2 IZt/d^2)$. Substitute for the first set of conditions putting $V = V_1$, etc.

$$\text{quantity of X radiation} \propto \frac{V_1^2 Z_1 I_1 t_1}{d_1^2}.$$

Similarly for the second set of conditions

$$\text{quantity of X radiation} \propto \frac{V_2^2 Z_2 I_2 t_2}{d_2^2}.$$

As the quantities of X radiation reaching the film in the two cases are equal,

$$\frac{V_2^2 Z_2 I_2 t_2}{d_2^2} = \frac{V_1^2 Z_1 I_1 t_1}{d_1^2}.$$

Therefore

$$I_2 t_2 = \frac{V_1^2 Z_1 d_2^2 I_1 t_1}{V_2^2 Z_2 d_1^2}.$$

But $V_1 = V_2$ and $Z_1 = Z_2$ and therefore the Vs and Zs cancel out in the equation.

Substituting the numerical values gives

$$I_2 t_2 = \frac{1^2 \; 15}{2^2} = 3{\cdot}75 \text{ mAs.}$$

The 'exposure' required under the new set of conditions is 3·75 mAs.

11 Thermionic emission and X-ray tubes

11.1 THE THERMIONIC DIODE

11.1.1 Thermionic emission. When an electric current flows in a conductor, 'free' electrons move along the conductor (4.1.1). The so-called 'free' electrons, however, are free only in the sense that they will flow in one of the two possible directions, and will behave in a manner described by **Ohm's Law** (4.2.2).

Electrons are light-weight particles (2.1.3) and therefore can move very rapidly under electric attraction or repulsion. Hence if they could be extracted from the relatively restrictive environment of a conductor (such as a copper wire) they might be made to perform more complex and useful functions than merely forming an electric current. This can be achieved in two ways: first, by expelling them into the free space in a vacuum, and second, by making use of them when they are situated between the regularly arranged atoms of a crystal of a material called a **semiconductor**. The latter method we shall discuss later (11.3.2); the former is achieved by the process of **thermionic emission**, as follows.

The atoms of a conductor, and with them the free electrons, are in a state of continuous vibration. The energy of the vibration determines the *temperature* of the conductor (1.4.1). At room temperature, both the atoms and their electrons attempt to escape from the surface of the conductor, but there are inter-atomic and electric forces which tend to hold them back. The net result is that very few atoms or electrons escape. If the temperature is increased, the vibrational energy of the atoms and electrons increases, until a certain temperature is reached when the most energetic of the electrons escape permanently from the conductor's surface. The latter process is called thermionic emission. As the temperature is further raised, even by just a small amount, more and more of the electrons escape. This process occurs at a temperature of around 2 000°C for pure tungsten (a metal widely used for this purpose in X-ray practice, 5.2.2); the effect is shown graphically in Fig. 11.1. Note that the electron emission increases *very rapidly* over quite a *small* temperature increase; this fact is important in the design of X-ray generators (13.5.1).

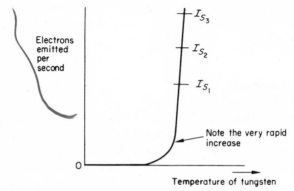

Fig. 11.1   The thermionic emission from tungsten.

    Thermionic emission can be regarded as a kind of 'boiling-off' or evaporation of electrons, although this process must not be confused with the actual boiling of the metal itself which would take place only at a very much higher temperature. In the latter case it is the tungsten *atoms* which would escape from the surface of the metal.

    The process of thermionic emission has enabled us to get electrons into free space; now we shall show how they can be made to perform more complex functions than when 'imprisoned' in a conductor.

11.1.2   The thermionic diode consists of a glass bulb from which practically all the air has been removed, and which contains a thin **filament** of tungsten (like the filament of an electric lamp, 5.2.2). Current is passed through the filament so that it glows white-hot, and electrons are emitted from it at a rate which is determined by the precise filament temperature. Close to the filament is situated a **plate** of metal whose shape depends on the particular application of the diode; it is usually called the **anode** but still retains the name 'plate' in the U.S.A. The filament and anode form the two electrodes which give the diode its name (di- = two; -ode = electrode); an electrode is simply a conductor by which electricity enters or leaves a volume of gas, liquid, etc. The filament is a particular example (universally used in X-ray generators) of the electrode whose general name is the **cathode**. Although the diode in general may be said to contain anode and *cathode,* we shall continue to speak of anode and *filament* because of the X-ray context.

Figure 11.2 shows a (thermionic) diode connected in a circuit whose sole

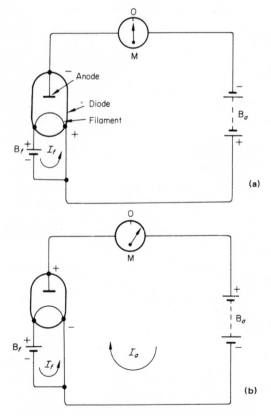

Fig. 11.2 The principle of the thermionic diode.

function is to demonstrate the properties of the diode; it has no other application. Notice the conventional symbol for the diode (4.1.2). The filament is heated by a current I_f flowing in the filament circuit from B_f, the filament battery. The anode is connected to the filament via a centre-zero moving-coil milliammeter M and an anode battery B_a whose direction can be reversed as shown in Figs 11.2a and b. Let us for the moment ignore the presence of the anode. The filament is heated to such a temperature that a moderate number of electrons leave it per second. What happens then? The electrons, having nowhere particular to go, form a cloud or charge in space, called the **space-charge** (shown in Fig. 11.2a). This, being negative, tends to repel any further electrons trying to escape from the filament, and when the space-charge has become sufficiently dense, the emission from the filament ceases.

Now consider the effect of the anode. If the anode battery B_a is connected so that the anode is *negative* with respect to the filament (Fig. 11.2a), the space-charge will be repelled back towards the filament. After the initial momentary movement, no further current flows. In this condition the diode behaves as a *non-conductor*. If, however, the anode battery is connected so that the anode is *positive* with respect to the filament (Fig. 11.2b), electrons are attracted from the space-charge; they reach the anode and flow around the external circuit to the filament, through the anode battery which provides the e.m.f. necessary to drive them around. At the same time, the depleted space-charge exerts less repulsive force at the filament and more electrons are emitted, thus replacing those lost to the anode. A continuous process ensues which results in a continuous flow of current around the anode circuit. This is called the **anode current** I_a. In this condition the diode behaves as a *conductor*.

It is evident that the diode has an important property which an ordinary conductor does not have. A conductor will allow electrons to flow equally in either direction, according to the direction of the p.d. across its ends. A diode, however, will allow electrons to flow *only* from filament to anode; this is because the filament end is hot and acts as a source of electrons whereas the anode end is cold and does not. This property of the diode is extremely valuable; when used in this way the diode is called a **rectifying valve** or **rectifier**. Its chief use, as we shall see later, is to convert alternating current into pulsating direct current.

11.1.3 The diode characteristic. In the last section the unique property of the diode, viz. its unidirectional conduction property, was explained in outline, and its application as a **rectifier** was mentioned. However, the diode has many other uses, the most important of which, in X-ray technology, is as an **X-ray tube**. The operation of this does not depend primarily on the *rectifying* action of the diode, but on other properties. It is necessary therefore to study the behaviour of the diode in greater detail; this may be done by reference to the demonstration circuit of Fig. 11.3. This circuit is an

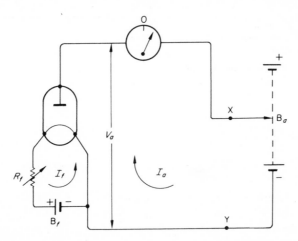

Fig. 11.3   A circuit to demonstrate the characteristic curve of a diode.

elaboration of that in Fig. 11.2, in that the filament current I_f can be varied by means of the rheostat R_f, and the anode-to-filament p.d. V_a can be varied (either positively or negatively) by a tapping on the battery B_a.

Suppose first that there is a moderate emission of electrons from the filament per second, and that V_a is made zero by connecting the tapping point X to point Y (thus excluding the battery B_a). The anode current I_a indicated on the meter M will be zero, because there is no e.m.f. in the circuit to drive electrons round. (This is only an approximation to the truth, but is sufficient for our present purpose.) This point is shown at the origin of the graph in Fig. 11.4. Such a graph is called the **characteristic curve** of the diode, or simply the **diode characteristic**.

Now suppose B_a is connected, but in reverse polarity, so that starting from zero the anode can be made progressively more and more negative with

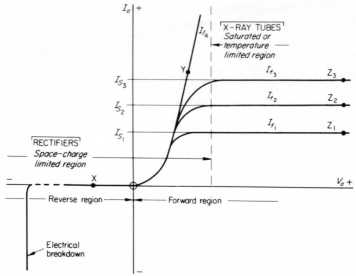

Fig. 11.4   The characteristic curve of a thermionic diode.

respect to the filament. I_a will still remain zero (11.1.2). This is shown as a horizontal 'curve' coincident with the $I_a = 0$ axis; it is called the **reverse region** of operation. If V_a is made more and more negative, however, thus travelling very far into the reverse region, a condition will be reached when there is sudden electrical breakdown, either *inside* the diode, due to ionization of residual gas molecules, or *outside* the diode, due to breakdown of the supporting insulators or to ionization of the air (3.3.2). The reverse region is a very important part of the diode characteristic, yet it is rarely illustrated explicitly in textbooks.

Now suppose B_a is connected with the polarity shown (Fig. 11.3) so that the anode is positive with respect to the filament, and that the p.d. V_a is gradually increased from zero. The curves show that whatever the value of I_f (within certain limits) the anode current I_a will increase, at first slowly, then more rapidly. This part is often called the **linear** part of the curve, although in practice it is not strictly linear. As V_a is further increased, the increase of I_a becomes slower, and a condition is eventually reached where no further increase in I_a is obtained, for a given value of I_f. This point marks the upper end of the **space-charge-limited region** (Fig. 11.4); the part beyond it is called the **saturated** or **temperature-limited region**. In this region, the anode is at such a high positive potential that all the electrons emitted by the filament at that temperature are attracted away from it immediately. In this region, the anode current I_a is independent of the value of the anode voltage V_a. If a change in I_a is required, it can be produced only by changing the rate at which the filament is emitting electrons, i.e. by changing the filament temperature by varying I_f. Because I_a is controlled only by the *filament temperature* this region is called *temperature-limited*. Fig. 11.4 shows characteristic curves for three different values of I_f: high (I_{f3}), medium (I_{f2}) and low (I_{f1}), corresponding to **saturation currents** (to the anode) of I_{s3}, I_{s2} and I_{s1} respectively. These currents correspond to values on the curve of Fig. 11.1 (11.1.1) and are marked on that curve for comparison. Notice that the three curves of Fig. 11.4 merge into one in the space-charge limited region. Also shown in Fig. 11.4 is a curve of I_a for a considerably greater filament current I_{f4} and therefore a much higher filament temperature. In this condition the emission from the filament (in electrons per second) is so copious that large currents may be passed by the diode at a relatively low value of V_a; this condition is employed in the rectifier (11.3.1).

We have discussed the diode characteristic in great detail because the diode is a very important device in X-ray technology. With necessary differences in mechanical design, an *X-ray valve* or *rectifier* is simply a diode operated in its *space-charge limited* region, and an *X-ray tube* is simply a diode operated in its *saturated* or *temperature-limited* region. In the remainder of this chapter we shall enlarge on the principles of this distinction and describe the practical details of construction of these devices.

11.2 X-RAY VALVES AND TUBES: PRINCIPLES

11.2.1 X-ray valves or rectifiers. Electrical power is distributed in most countries, via the supply mains, in the form of **alternating current** (Chapter 8). However, many devices, for example X-ray tubes, do not function well with alternating e.m.f.s applied; it is therefore necessary to convert the *alternating* e.m.f. or current to a *unidirectional* e.m.f. or current (8.1.5). This is done by the device known as a **rectifier**; in X-ray technology it is commonly called an **X-ray valve**, or simply **valve**.

Figures 11.5a and b illustrate the principle of a rectifier. In Fig. 11.5a, the source of alternating e.m.f. drives an alternating current through the device represented by R; we shall assume that R functions best with unidirectional

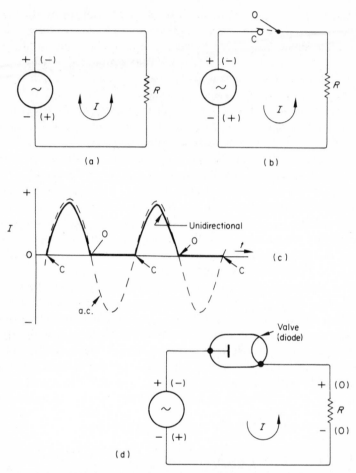

Fig. 11.5   The principle of a half-wave rectifier; the arrows indicate the flow of electrons.

current. In Fig. 11.5b, a switch has been interposed between the a.c. source and R, and it is assumed that some supernatural agency is available to open and close the switch at exactly the correct instants so as to *'let through'* the *positive* half-cycle but to *stop* the *negative* half-cycle. The result of this imagined behaviour is shown in Fig. 11.5c; the dashed curve shows the alternating current which would flow in Fig. 11.5a and the full curve shows the unidirectional current of Fig. 11.5b. For this imaginary arrangement to work the switch would not only have to operate at *exactly* the correct instants (show by O = open, C = closed in Fig. 11.5c), but also it would need to operate one hundred times per second (for a 50 Hz supply). Some early X-ray generators did in fact have switch-rectifiers, but they were unreliable and unsuccessful. We shall now show that a simple diode will fulfil the function of the switch automatically and reliably.

In Fig. 11.5b, the switch, when *closed,* has a resistance of *almost* zero; when *open* its resistance is *nearly infinite* (∞). Now let us estimate what is the resistance of the diode whose characteristic is shown in Fig. 11.4 (11.1.3). At point X, in the reverse region, the resistance $R = V/I = V/\text{zero} = \infty$. (Any finite number divided by zero is infinity!) At point Y, in the forward region, the current is moderately high, the voltage moderately low, hence $R = \text{low/high} = \text{low}$. Hence the diode, so long as it is operated in the space-charge-limited region, will behave in the same way as the switch in Fig. 11.5b, offering only a *low* resistance to the *forward* e.m.f. but an *almost infinite* resistance to the *reverse* e.m.f. Moreover, its resistance changes automatically according to the direction of the p.d. across it. The circuit of Fig. 11.5d shows how a single diode can be used to produce a *half-wave* unidirectional flow of current from an alternating source of e.m.f. This is therefore called a *half-wave rectifier* circuit. In this circuit the diode behaves very similarly to a motor car tyre 'valve', which allows air to enter the tyre from the pump but prevents it from leaving the tyre. The tyre valve is in fact operated by the air 'pressure difference' (p.d.) across it, analogous to the electrical p.d. across the diode. We shall discuss practical aspects of rectifiers in sections 11.3.1 and 11.3.2, and practical circuit arrangements in Chapter 12.

11.2.2 X-ray tubes. X rays are produced whenever high-energy electrons are strongly decelerated (caused to lose their velocity rapidly) (10.2). This may be simply achieved in the diode by applying a very high p.d. between anode and filament, resulting in the electrons from the filament acquiring a very high velocity and therefore a high energy by the time they reach the anode. When the electrons strike the anode they are decelerated (10.3) and produce X rays; the properties of the X rays and the factors influencing these properties were described in Chapter 10. In this section we shall discuss the electrical conditions necessary for efficient X-ray production.

For clinical use, the X rays must be sufficiently energetic to penetrate large thicknesses of matter, e.g. soft tissue and bone. This can be achieved in an X-ray tube only by applying between 40 and 250 kVp between target and filament. If the X-ray tube were operated in the *space-charge-limited* region, as is the X-ray *valve* or rectifier (11.2.1), the diode resistance (on the continuation of the curve for I_{f4} to high values of V_a in Fig. 11.4 (11.1.3)) would be low and a very large current would flow. This would have many harmful effects: it would damage the X-ray tube, it would damage the generator circuit supplying the X-ray tube, and it would produce too large an intensity of X rays. Hence steps must be taken to limit the current, preferably by increasing the resistance of the X-ray tube itself. This can be most easily achieved by operating the tube as a diode in the *saturated* or *temperature-limited* condition, by reducing its filament temperature (11.1.3). For example, it can be seen that the diode resistances at points Z_1, Z_2 and Z_3 (Fig. 11.4) can be made larger than in the space-charge-limited region, because the target-filament p.d. can be made as large as necessary, independently of the value of I_a ($R = V/I$). The high value of resistance is simply achieved by regulating the value of I_a by reducing the filament current and hence the filament temperature (11.1.3).

A secondary but very important advantage of this mode of operation is that the anode voltage and current, called 'kV' and 'mA' respectively in radiological practice, are almost entirely independent of each other. Thus the kV can be set to give the best depth dose or contrast for the particular subject (14.6.2, 14.6.3), then the mA adjusted independently to produce the required X-ray exposure rate (10.5.3).

We have shown in sections 11.2.1 and 11.2.2 how the diode, operated in different conditions, can be used both as an X-ray valve (rectifier) and as an X-ray tube. However, the reader must not imagine that the *same* diode can be used for both purposes. The two functions require quite different types of construction; these will be discussed in sections 11.3.1 and 11.3.3. Nevertheless, *in principle,* both devices behave electrically as diodes and have a common 'ancestry'.

11.2.3 The cold-cathode gas-filled diode. The inclusion of a small amount of an inert gas such as neon or argon (to a pressure of about 10 mm of mercury) in the diode completely changes its characteristics. It is then known as a **gas-filled diode**, and the filament is replaced by an unheated cathode. It is sometimes therefore called a **cold-cathode diode**. Its properties may be demonstrated by means of the circuit shown in Fig. 11.6. (Note the conventional symbol including the cold cathode and the 'gas filling'.)

Suppose we start with the battery voltage V_B at zero; the anode current likewise will be zero. Now let us increase the battery voltage slowly; the current increase will be negligible because the electron emission from the cold cathode is negligible, although there is a very small emission at room temperature. The electrons which *are* emitted will very soon strike a gas molecule; because their energy is low, no ionization will take place. As V_B is progressively increased, a stage will be reached when the energy of these few

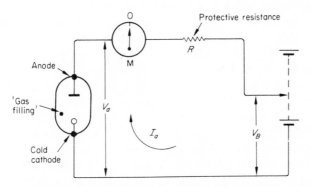

Fig. 11.6   The principle of the cold-cathode gas-filled diode.

electrons is sufficient to ionize the gas molecules; this results in the expulsion from the molecules of more electrons, which in turn cause further ionization. The net result is a very large 'burst' of ionization (cumulative ionization), from which electrons travel to the anode producing a very large increase in anode current. This corresponds to a sudden decrease in the resistance of the diode at a certain value of V_B; if it were not for the presence of the protective resistance R a damagingly high current would flow.

The gas-filled diode differs from the thermionic diode in that its conduction takes place suddenly at a particular value of anode voltage, whereas in the thermionic diode the anode current increases gradually as the anode voltage is increased. The gas-filled diode is like a switch which is either 'on' or 'off'; once it is switched 'on' (i.e. current is flowing) the only way to stop the current is to reduce V_a to zero. The gas-filled diode thus acts like an electrical 'trigger', which allows a changing voltage (V_B) to initiate some electrical event (e.g. the end of an X-ray exposure) at an appropriate instant. This type of application will be briefly discussed in Chapter 13.

11.3   X-RAY VALVES AND TUBES: PRACTICAL ASPECTS

11.3.1   X-ray valves or rectifiers. The principle of the rectifier was discussed in section 11.2.1. The practical requirements for rectifier construction are as follows.

(i) The insulation between anode and filament should be very good so as to achieve an external reverse resistance as high as possible. There should also be very few gas molecules left in the valve so that cumulative ionization (11.2.3) is unlikely to occur even with very high reverse voltages.

(ii) There should be a copious emission of electrons from the filament so that the resistance of the diode in the forward direction is low (11.2.1; Fig. 11.4 (11.1.3), point Y, I_{f4}). This ensures that the voltage drop across the rectifier is reasonably small for a given current in the forward direction.

These requirements are best satisfied by a diode consisting of the usual evacuated glass bulb, in which the filament is a long tungsten wire formed into a number of loops. The anode is usually a cup-shaped metal electrode surrounding the filament. Because the voltage drop across the rectifier is small in the forward direction, the electrons arriving at the anode have quite low energies (of the order of 1 keV) and the power dissipated at the anode in the form of heat ($P = VI$) is small. The anode can therefore lose its heat easily by radiation (1.4.3) and no special difficulties arise as they do in X-ray tube design (11.3.3). It is usual to immerse the whole rectifier in oil so that the exterior resistance and the dielectric strength (3.4.2) are high. The condition of electrical breakdown (Fig. 11.4, 11.1.3) is thus less easily reached.

11.3.2   Solid-state (selenium and silicon) diodes. An alternative to the use of electrons in a vacuum, as in the thermionic diode, is to confine them between the regularly arranged atoms of a **semiconductor** (11.1.1). A semiconductor is a crystalline material of a special kind that is composed of an element, such as selenium or silicon, having four electrons in its outer shell (2.2.2). Electron shells are particularly stable when complete, in this case when containing eight electrons. A crystal of a semiconductor contains atoms that 'share' each other's outer electrons. Every atom is surrounded by four other atoms, each of which shares one of its outer electrons with the central atom. The result is an exceedingly stable material which has (in principle) no free electrons, and which is therefore a very bad conductor.

However, such a material can be made to conduct by introducing into it layers of impurities. If the impurity atom has five electrons in its outer shell, it can *supply* free electrons and is called an **N-type** (negative) impurity; if it has three electrons it can *accept* free electrons and is called a **P-type** (positive) impurity. Without going further into solid-state theory it can be seen in a general way that if N-type and P-type semiconductors are placed in contact, there will be a preferential direction of electron flow from the N-region to the P-region. Thus the semiconductor will act as a rectifier. It has the outstanding advantages that it is relatively small and, because it needs no filament supply, it is much more reliable and can lead to great circuit simplification. These advantages are of importance in the design of X-ray generators (Chapter 13).

Solid-state diodes have been available for low voltages (up to a few hundred volts) for many years; an important low-voltage application is in the 'bridge' rectifier used to enable alternating currents to be read on a moving-coil meter (12.4.1). However, with the development of improved semiconductor materials from 1960 onwards, solid-state diodes have become available for high-kV use in X-ray generators. Details will be found in the latest books on X-ray apparatus construction.

11.3.3 X-ray tube design is a complex subject. The final result is a compromise between many conflicting factors, some of which are not concerned at all with electrical properties. Those aspects of X-ray tube design concerned with the production of radiation were discussed in Chapter 10. For the present purpose, it is sufficient to bear in mind that electrons from the filament strike the target with high energy, from 10 to 250 keV, depending on the application, and that X-rays are emitted in all directions. The area of the target struck by the electrons, which therefore forms the source of the X rays, is called the **focal area**. The part of the anode on which the focal area is formed is made of tungsten (10.2.2); in a conventional X-ray tube (Fig. 10.1, 10.2.2) this tungsten disk is usually embedded in a larger block of copper known as the **anode block**.

Most of the electrical energy imparted to the electrons as kinetic energy (1.3.5) is converted into heat in the focal area; less than 1% of the energy appears as X rays. The X-ray tube designer's greatest problem is to arrange for this heat to be removed as rapidly as possible; he must ensure that the production of X rays proceeds efficiently despite the large increase of temperature which results even when a large part of the heat is removed.

Let us first discuss the X-ray requirements, disregarding the heat problem. Electrons travel to the target from the filament, and to ensure unimpeded emission of X rays it would seem natural to make the target surface at an

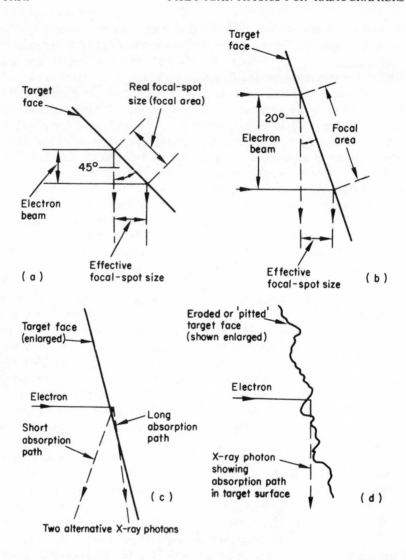

Fig. 11.7   The relation between real and effective focal spot sizes; self-absorption in a target.

angle of 45 degrees relative to the electron direction. The centre-line of the X-ray beam (called the X-ray beam **axis**) would also be at 45 degrees to the target surface. This condition is shown in Fig. 11.7a, from which it will be seen that the actual or **real size** of the focal area on the target surface is not the same as the electron beam cross-section. If the latter is assumed square and, for example, of 1 mm side, the real size of the focal area will be 1 mm wide and $1 \times \sqrt{2} = 1.4$ mm long (because $1/\sin 45° = \sqrt{2}$). However, if an observer, capable of seeing X rays with the unaided eye, were to look along the X-ray beam axis he would see a focal spot of apparent or **effective size** 1 mm wide and 1 mm long. (It is, of course, *most dangerous* ever to look into an X-ray beam!)

In most applications of X-ray tubes it is important to have a small effective size of focal spot. In radiography and fluoroscopy, which use what are virtually 'shadow' pictures, the *sharpest* picture will be obtained with the geometrically *smallest* radiation source (14.7.2). In radiotherapy, the treated area is delineated by a shadow cast by the treatment applicator, and the sharpest edge to the beam will be produced by the smallest focal spot, although this requirement is not so stringent as in radiography. Diagnostic radiology requires effective focal spot sizes of 0.3 to 2 mm square (with 0.1° and 6 mm as outside limits for special applications) while radiotherapy demands no better than 10 mm. Why should we not have the smallest possible focal spot for all applications? The reason is two-fold: first, that it is difficult to focus electron beams on to very small areas, second, that if all the electrons *could* be concentrated into a small area, great difficulty would be experienced in getting rid of the very large concentration of heat so produced. In other words, for ease of heat removal we require a *large real* focal area, and for good radiological sharpness a *small effective* focal spot. This is achieved by making the surface of the target at a small angle to the X-ray beam axis. Fig. 11.7b, for example, shows a **target angle** of 20 degrees; at this value, for an *effective* size of focal spot of 1 mm × 1 mm the *real* focal area is 3 mm × 1 mm. This is sometimes called a **line focus**. The heat escapes by conduction much more readily from the large focal area, and it is easier to focus the electrons on an area of this size.

In section 10.2 we described how X rays are *produced* in a target. However, the X rays produced, to be of any use, must *emerge* from the target; this is a requirement which has many implications in target design and use. First, it limits the extent to which the effective focal spot size can be reduced by decreasing the target angle. Fig. 11.7c shows how Bremsstrahlung produced in the first two or three molecular layers of the target surface can emerge more easily towards the filament end of the tube (1) than towards the anode end (2), because in the latter direction the X rays must traverse a longer path in the tungsten than in the former direction. This is known as the 'anode heel' effect; it results in the exposure rate at the anode side of the X-ray beam being smaller than that at the filament side. The inequality affects both radiography and radiotherapy but is usually more important in the latter. Second, if the focal area on the target is rough or 'pitted', because of either faulty manufacture or target erosion (11.3.4), X rays produced at the bottom of the 'pits' may have to traverse a longer path in tungsten before emerging in any direction and may thus be considerably attenuated. This condition is shown in Fig. 11.7d.

Figure 11.8 shows the salient features of the filament-target assembly of an X-ray tube. The filament consists of a spiral of thick tungsten wire operated from a low voltage, high current source. The filament emits electrons in all directions; to focus these into a rectangular area on the target,

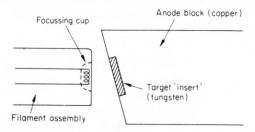

Fig. 11.8   The salient features of a conventional X-ray tube filament–anode assembly.

the filament is partially enclosed in a semi-cylindrical metal cup, called the focussing cup. This, together with the nearby target, produces an electric field (3.1.2) of such a shape that the electrons are roughly **focussed** on to the desired target area. The focussing cup also serves to prevent electrons from the filament from charging and perhaps damaging the glass wall of the tube.

Sometimes it is necessary to make a rough measurement of the effective size of the focal spot. This may be done by taking a non-screen radiograph (i.e. with film alone to give high definition) of a special kind of 'pinhole' made in a block of lead or similar metal. If, for example, the distances from the focus to the pinhole and from the pinhole to the film are equal, then the size of the image on the film (after development) will be equal to the effective focal spot size (Fig. 11.9). However, there are many complicating factors in this measurement and the result obtained should be regarded as only very approximate.

11.3.4   X-ray tube cooling. One reason for using tungsten for the target material is that it is a fairly good conductor of heat; the heat produced by the electrons therefore escapes fairly readily into the main copper anode block. The precise means adopted for cooling the tube, however, depend on its application. The types of medical application are summarized in Table 11.1.

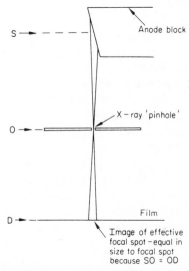

Fig. 11.9 (11.3.3) The approximate measurement of effective focal spot size using a pinhole.

In both fluoroscopy and radiotherapy, the power involved, that is the *rate* of production of heat in the tungsten target, is relatively small. The heat is easily conducted to the copper anode block, and from there it is removed in one of two ways. All X-ray tubes are enclosed in metal cases which are both shockproof (12.3.3) and X-ray proof, the cases being filled with oil for insulation purposes. For fluoroscopy and superficial therapy, the heat arriving in the copper block is removed sufficiently quickly by oil convection (compare the transformer, 8.3.2) and then from the metal case by air convection. Sometimes a small fan is arranged to blow air on the case to

159

TABLE 11.1 (11.3.4)   *Typical conditions of use of medical X-ray tubes*

Time factor	Use	Typical factors			Power	Exposure time	Energy per exposure
Continuous	Fluoroscopy	100 kV*	0·5 mA		50 W	300 s	15 kJ
	Radiotherapy	250 kV*	15 mA		3·75 kW	600 s	2 250 kJ
Intermittent	Radiography	100 kV*	500 mA		50 kW	0·2 s	10 kJ

* The p.d. across the tube is for simplicity assumed constant, although this is by no means always true (Chapter 12).

assist cooling. Such a tube is shown in Fig. 10.1 (10.2.2). In so-called 'deep' radiotherapy, however, when the total energy per exposure is very large (Table 11.1) (but spread over a long period), it is necessary to remove heat from the copper block by more positive means. This is done by pumping oil in and out of a cavity in the rear of the anode block (Fig. 11.10a). The oil is in turn cooled in a heat exchanger by cold water from the mains supply.

High-power radiography, however, poses quite a different problem. Here the total energy per exposure (Table 11.1) is quite small; it is therefore unnecessary to make any special provision for *continuous* cooling of the anode block. The greatest difficulty however is caused by the *rate of production* of heat in the target, viz. the power, which Table 11.1 shows might be as much as 50 kW; it can be even more in a very large X-ray set. At this power the heat is no longer able to flow by conduction from the tungsten to the copper sufficiently fast to prevent a harmful rise in temperature of the focal area. For many years this limited the maximum radiographic exposure possible, until the invention in 1933 of the **rotating-anode X-ray tube**. The principle of this tube is illustrated in Fig. 11.10b; it effectively incorporates means by which the ratio of the *real* focal area to the *effective* size of the focal spot can be increased far above the value determined by the angle of the target (11.3.3). This is done by making the anode in the form of a tungsten disk with its edge bevelled to the correct angle. This disk is mounted on a shaft which is driven by an **induction motor** (6.3.3 and below). Before the radiographic exposure is made, the motor is energized, the anode disk rotates very rapidly, and the electrons that generate the X rays thus strike a focal area that extends over its whole circumference. No part of the anode therefore attains a damagingly high temperature; the effective focal spot size

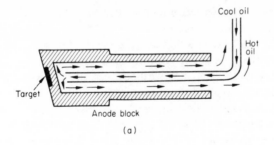

(a)

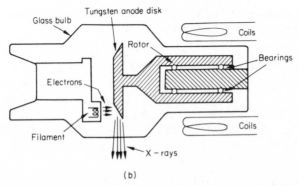

(b)

Fig. 11.10   Designs of X-ray tube for high-power radiotherapy and radiography: (a) the anode block of a 'deep' therapy tube cooled by oil, and (b) a rotating-anode tube for radiography.

is unaltered by this artifice. Heat is removed from the anode disk of the rotating-anode tube mainly by radiation. A recent development in anode design is to make part of the anode disk of hard carbon, which has a high specific heat (1.4.2) compared with tungsten. Thus the amount of energy permissible in the exposure for a given temperature rise of the anode is even greater.

The rotating anode is driven by an induction motor. On the shaft of the anode disk is mounted a cylinder of copper, called the **rotor**. Outside the X-ray tube (Fig. 11.10b) are coils of wire through which an alternating current is passed when rotation is required. The current produces an alternating magnetic field which induces an alternating e.m.f. and hence a current in the rotor; the magnetic field due to the latter current interacts with that of the external current to produce rotation. The use of this type of motor, requiring no electrical connexions to be made to its rotor, simplifies the design of the tube. The bearings carrying the rotating shaft are lubricated with a dry lubricant and normally last the whole life of the tube.

Despite the use of these cooling methods, occasionally X-ray tube targets overheat because of a fault in the apparatus or because of persistent tube overloading (11.3.5). What happens then? The result of the overheating is that the surface layers of the target evaporate, leaving it with an eroded or 'pitted' surface. This is not only a less efficient X-ray source (because of self-absorption, 11.3.3) but also the evaporated tungsten tends to be deposited on the inside of the glass tube thus increasing its inherent filtration (14.5) and reducing the tube output. The tungsten deposit may also cause irregular conduction paths within the tube, resulting in electrical instability and breakdown.

11.3.5 X-ray tube ratings. When an X-ray tube is installed for a particular group of applications, for example for fluoroscopy and radiography, it may be assumed that it is suitably designed in accordance with principles that have been briefly outlined in sections 11.3.3 and 11.3.4. It then remains only to *use* it in a manner that will give the best possible radiological results and that will ensure a long life for the tube; this is partly the responsibility of the radiographer.

To ensure proper use, two types of requirement in general need to be satisfied. First, the X-ray tube should be appropriately positioned in relation to the patient and the film, and suitable exposure factors (kV, mA and time) chosen, so as to give the desired result. Second, the factors chosen should not be such that the tube is overloaded, causing rapid deterioration and perhaps early breakdown.

Tube positioning, etc., is studied elsewhere under the headings of radiographic and radiotherapeutic techniques. Tube overloading is avoided by the use of certain graphical and other information supplied with the tube and given the general name **X-ray tube ratings**. The subject of tube rating is complex and will be studied elsewhere under the heading of apparatus construction; here we shall merely summarize the basic physical concepts that are relevant.

X-ray tube rating dictates firstly, what maximum combinations of kV, mA and time may be used, and secondly, *how often* these may be used in a given total period of time. Two types of limitation exist: (i) electrical and (ii) thermal.

The maximum permissible (peak) kV, regardless of mA, is determined by the design of the tube and of its shockproof housing (12.3.3). The kV must not be so great as to exceed the dielectric strength of the insulation *outside* the tube or to ionize the small amount of residual air *inside* the tube. The maximum permissible mA, regardless of kV, primarily depends on the tube filament design and its maximum available electron emission, although certain secondary effects associated with space-charge formation complicate the situation. These are both *electrical* limitations.

The maximum permissible kV and mA considered *together,* however, are determined by the maximum power ($P = VI$) that may be dissipated in the focal area; this is primarily a *thermal* limitation. Depending on the tube application, it may or may not be associated with exposure time. For example, a 'deep' therapy tube, because it is continuously cooled, may be operated almost indefinitely at, say, 200 kV and 15 mA. On the other hand, the rating of a rotating-anode tube, which depends on the thermal capacity of the anode disk and on its subsequent cooling by thermal radiation, involves the exposure time.

In X-ray technology it is usual to express energy in terms, not of joules or calories, but of **heat units** (H.U.). The number of heat units involved in a typical radiographic exposure, for example, is given by

$$\text{number of heat units} = \text{kVp} \times \text{mA} \times \text{seconds}.$$

The relationship between heat units and joules cannot easily be derived from this equation because the kV waveform must be taken into account.

The maximum energy permissible in a single exposure is expressed in heat units. However, this exposure would raise the focal area to its maximum permissible temperature; in principle, one would then have to wait a very long time for the target to cool down sufficiently for a second similar exposure. In practice, the normal interval between one patient and the next is an adequate cooling time when single exposures are concerned. However, for multiple exposures, e.g. in serial angiography, exposures may be made at a faster rate provided a lower energy (kV.mA.s) per exposure is used. The X-ray tube ratings give all the required information for every possible routine application. However, in most modern X-ray generators there are circuits that automatically ensure that the X-ray tube ratings are not exceeded.

12 High-voltage rectifier circuits

12.1 INTRODUCTION

We have now described sufficient basic physics to be able to explain the design and operation of X-ray generators. We shall do this in two parts. First, the X-ray tube, for the most efficient operation, requires a unidirectional (8.1.5) voltage of 50 to 250 kV. This high voltage is produced from the mains supply by a voltage step-up transformer (8.2.2), called the high-voltage (high-tension) transformer, followed by the conversion of the alternating voltage to unidirectional by one or more rectifiers, which may be either thermionic diodes (X-ray valves, 11.2.1, 11.3.1) or solid-state (semi-conductor) diodes (11.3.2). These are arranged in one of several possible patterns of circuit, according to the application. The transformer and rectifiers together are described as the **high-voltage rectifier circuit**; its different forms will be explained in this Chapter. Second, the high-voltage rectifier circuit itself must be included as a unit in a more complex circuit that embodies devices for controlling and indicating the values of various factors, e.g. kV, mA and exposure time, as well as other devices for performing subsidiary functions such as stabilizing voltages. This more complex circuit is described under the general term of the **control and indicating circuits**; a simplified version of the control and indicating circuits will be discussed in Chapter 13.

12.2 SELF-RECTIFYING CIRCUITS

12.2.1 The self-rectifying circuit. If we recall that one of the requirements for the production of X rays is a unidirectional flow of electrons from the

164

filament to the target, accelerated by an adequate value of p.d., we may conclude that it is sufficient to connect across the X-ray tube an alternating voltage of the correct peak value, as in Fig. 12.1a. In this circuit, the X-ray tube acts as its own rectifier, hence the name **self-rectifying circuit**; no additional valves are required. (The reason for the earth connexion to the centre point of the transformer secondary will be explained in sections 12.3.2 and 12.3.3).

Figure 12.1b shows the voltage and current waveforms in this circuit. Both half-cycles of the alternating voltage appear across the tube; thus if the peak positive voltage on the target (with respect to the filament) is, say, 70 kVp, then during the opposite half-cycle (the polarity shown in brackets) a negative peak of 70 kVp will appear at the target. The positive (forward)

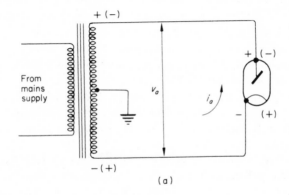

(a)

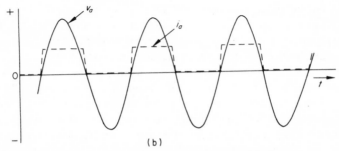

(b)

Fig. 12.1   The self-rectifying circuit, with waveforms.

voltage half-cycle will drive current through the tube, thus producing X rays, whereas the negative (reverse) voltage half-cycle will not. The waveform of the tube current will *not* be a half-sinusoid; because a relatively small target voltage causes the tube to reach saturation (Fig. 11.4, 11.1.3), the current waveform is almost rectangular (shown dashed in Fig. 12.1b).

The self-rectifying circuit is widely used for *low-power* X-ray generators such as dental units. In these, the high-voltage transformer and the X-ray tube are usually enclosed in the same oil-filled container, resulting in a very simple arrangement. However, the circuit is not suitable for *high-power* generators for reasons given in the next section.

12.2.2 The disadvantage of self-rectification. The ability of a diode (in this case the X-ray tube) to allow electron flow in only one direction depends on the fact that one electrode is hot and the other cold; this gives the diode its asymmetrical characteristics (11.1.2). However, if an X-ray tube, operated from *any* circuit, is used for high-power radiography or radiotherapy, viz. at high kV and high mA ($P = VI$), the large dissipation of power at the target will cause it to become very hot (despite cooling measures, 11.3.4). It may even be hot enough to emit electrons, though not so hot as to cause focal-area damage. This condition is not *in itself* harmful to the tube, so long as the melting point of the target material is not reached even in localized regions of the focal area (11.3.3, 11.3.4).

However, if this condition occurs with a *self-rectifying* circuit, serious damage to the tube will result, probably culminating in its destruction. This is because the filament becomes positive with respect to the target during the reverse (negative) voltage half-cycle; the electrons emitted from the hot target are then accelerated across the tube and impinge on the filament with high energy. But the filament is a thin wire of small thermal capacity (1.4.2) and is already white hot. Hence the rise in temperature resulting from the electron bombardment will almost certainly cause the filament to melt and collapse.

For this damage to occur, *two* conditions must arise simultaneously: first, high-power exposures, resulting in a considerable rise of temperature and electron emission from the target, and second, the application of the reverse voltage across the tube. Removal of *either one* of these conditions will avoid the damage. Thus, we have already seen that the self-rectifying circuit is suitable for *low-power* exposures. However, when high-power exposures are required, it is necessary to ensure that the reverse voltage is prevented from reaching the tube. This is achieved by the use of separate *rectifiers* or *valves*.

12.3 HALF-WAVE PULSATING-VOLTAGE CIRCUITS

12.3.1 The two-valve half-wave circuit. The principle of the diode used as a half-wave rectifier was explained in section 11.2.1. In particular, Fig. 11.5d shows how a valve (diode) may be used to ensure that the reverse voltage from the alternator does not appear across the resistance R. Fig. 11.5c illustrates the disappearance of the reverse half-cycle; during this interval the voltage across R is zero. If, in the Figure, the alternator is replaced by a high-voltage transformer, and R by an X-ray tube, we have a simple half-wave rectifier circuit which is identical to the self-rectifying circuit of Fig. 12.1 (12.2.1), except for the addition of the rectifier and the absence of the earth connexion (12.3.2).

Although the circuit of Fig. 12.1, with the above modifications, would function quite well as a half-wave X-ray generator, certain additions must be made to make it effective and reliable in practice. These additions consist of an extra valve and the central earth connexion (Fig. 12.2a). The extra valve is necessary for two reasons. First, if there were only a single valve in the circuit, and a breakdown were to occur in the valve (such as cumulative ionization (11.2.3) or a mechanical short-circuit) so that its reverse resistance became low (11.2.1), the reverse voltage from the transformer would be applied across the X-ray tube, with the harmful consequences described in section 12.2.2. The presence of a second valve, in series with the first, ensures a 'second line of defence'. Second, it is customary to make X-ray generator circuits **symmetrical** (or *balanced*) about earth in respect of voltage. This accounts for the centre earthed point, and provides an additional reason for the presence of the second valve. The circuit, of course, could not be symmetrical with only one valve. The reason for the symmetry will be explained below (12.3.2, 12.3.3).

Figure 12.2b shows the waveforms in the circuit. During the 'bracketed' half-cycle of v_a (shown in Fig. 12.2a), the rectifiers are non-conducting; hence the tube voltage v_a is zero, and no acceleration of electrons from target to filament can take place. During the other (forward) half-cycle, the two rectifiers and the X-ray tube conduct, and saturation current (in the X-ray tube) flows for the greater part of the half-cycle. The resistance of the rectifiers is kept low by ensuring a copious electron emission from their .filaments, thus enabling them to operate in the space-charge limited region (11.3.1).

The two-valve half-wave circuit is used for certain 'superficial' therapy

167

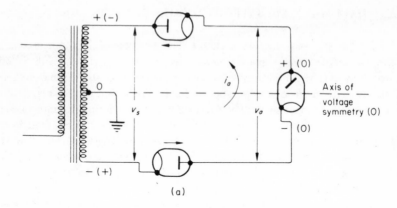

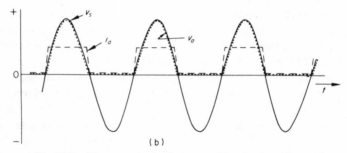

Fig. 12.2   The two-valve half-wave rectifier circuit, with waveforms.

generators operating up to 120 kV at 7 to 15 mA. Before describing more advanced generators, we shall explore the reasons for symmetry in X-ray generator circuits; these reasons are complex and bound up with electrical hazards and safety requirements.

12.3.2　Electrical hazards and precautions. The use of any X-ray apparatus is accompanied by hazards of three kinds: radiation, fire and electric shock. Radiation hazards are discussed in section 12.3.3; fire is avoided by the use of fuses (5.2.2) and insulation of adequate quality. Electric shock is caused when a voltage (p.d.) is applied between two or more points on the human body thus causing a current to flow through the body. The effect of a medium-sized current flow is to cause a disturbance in muscle function; this often appears as a sudden muscular contraction and the victim is often 'thrown' away from the offending contact (although he himself does the throwing!) Larger currents may cause burning of the skin.

Whether or not an electric shock is dangerous depends on the path of the current flow in the body. For example, a current flowing between finger and thumb of one hand may be unpleasant and even disfiguring but will not be lethal. However, if the current flows through the trunk, for example from one hand, along the arm and down to the legs and feet, it can arrest normal breathing and heart action, resulting in death. This type of shock most commonly occurs if the operator of the electrical equipment is standing on a wet or otherwise conducting floor and touches an uninsulated high-voltage point. In unfavourable conditions, e.g. with wet hands, voltages lower than 200 V can be lethal. It is therefore essential to design electrical equipment to ensure that this can never happen.

There are two general methods of making electrical apparatus safe. The first, which is limited to small domestic units such as hairdriers, is to enclose the electrical components in a container made of a double layer of insulating plastic. Then if one layer of insulation should develop a fault, there will be a second 'line of defence'.

The second method is used for all large units, domestic and otherwise. It consists of enclosing the electrical components in an earthed metal case which is insulated internally from the high-voltage connexions. Fig. 12.3a shows diagrammatically the electrical components EC of an X-ray generator, supported inside a metal case MC by insulators I. Assume initially that the metal case is *not* earthed. One of the insulators has developed a fault which causes it to conduct electricity. The electrical components are connected to the mains supply, of which one connexion, the **neutral main** N, is connected indirectly to earth, while the **live main** L is at about 240 V above earth and is assumed to be driving a 'leakage' current through the insulator fault F, the metal case and the operator O. The operator is touching the case (e.g. leaning on it) and is assumed to be standing on a conducting floor connected to earth E. The current returns to the mains supply through earth to the neutral main. This situation is obviously highly dangerous to the operator, and may even be lethal. It is avoided, and the whole system rendered completely safe, by the simple expedient of *securely* earthing the metal case. Fig. 12.3b shows that this mode of connexion provides an alternative path for the leakage current. One of the two alternative paths is of very low resistance through the earth connexion, the other is of much higher resistance through the operator. The

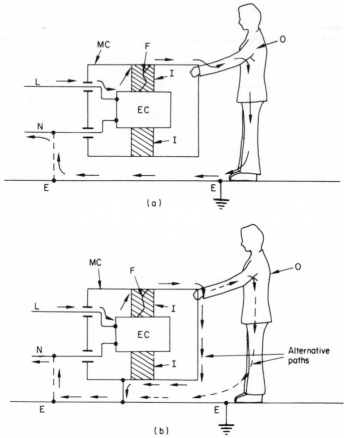

Fig. 12.3   The safe construction of electrical equipment.

current therefore divides between these paths in *inverse* proportion to their resistances, i.e. practically none of the current flows through the operator (compare currents through meter and shunt, 4.3.2).

Every part of all modern X-ray units is protected in this way, though the presence of the metal case is not always obvious, for example in high-tension cables (12.3.3). It must be strongly emphasized that *such equipment is safe only if the metal case is firmly earthed via a low resistance link*. If a radiographer at any time has any reason to think that the earthing of an X-ray set may be faulty, for example because an electric shock has been felt, some competent person, e.g. an engineer or physicist, *should be informed immediately.* This is most likely to happen with mobile equipment in which the earth link is made via the third pin of a plug and wall socket.

12.3.3 X-ray cables and shields; generator symmetry. The X-ray tube in its shield, together with the **high-tension cables** leading to it from the generator, are a good example of the earthed metal case technique (12.3.2). The tube is enclosed in a steel case (the shield) of appropriate shape, filled with oil for insulation and cooling purposes (11.3.4). Radiation hazards are reduced to a minimum by lining the X-ray tube shield with lead except for the apertures for the entry of cables and oil-tubes and for the emergence of the radiation itself. The X-ray beam is further confined to the minimum extent compatible with usefulness by diaphragm boxes or cones, and rules should be observed about the use of the beam in relation to the patient and operators (17.3.2, 17.3.3).

Current at high voltage is led to and from the X-ray tube by flexible high-tension (high-voltage) cables. These consist of an inner conductor surrounded by a thick layer of insulating rubber of high dielectric strength (3.4.2). The rubber is in turn encased in a tube of flexible material (called 'braid') woven from thin copper wire; this acts as a flexible metal case. The whole is covered by a fabric or plastic outer layer. Both the tube shield and the metal braids of the high-tension cables are securely earthed.

We can now see why it is desirable for an X-ray generator circuit to be symmetrical with respect to earth. Fig. 12.4a shows diagrammatically an X-ray tube and cables, connected to a generator G (represented by a box) which is earthed at the end connected to the X-ray tube filament. For simplicity, the generator is assumed to produce a constant voltage (12.5.2) of 100 kV and the potentials of relevant points in the circuit are shown relative to earth as zero. Both the filament connexion and the metal case at the filament end are at zero potential, hence the insulation here has zero potential difference across it. On the other hand, the target connexion is at +100 kV, and the insulation at the target end of the circuit must withstand this full value. The insulation in the set is very unevenly stressed.

Figure 12.4b shows the effect of earthing the generator centre point. The target and filament connexions are at +50 kV and −50 kV, respectively, therefore the insulation is equally stressed and, just as important, no

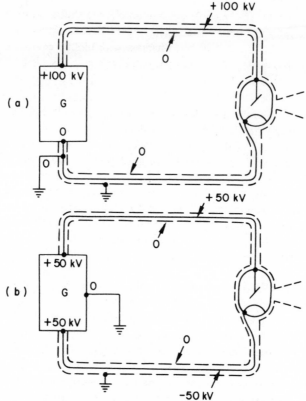

Fig. 12.4   The reason for the symmetry of X-ray generators.

insulation need withstand a p.d. of more than 50 kV. This symmetry makes the practical design of the generator much easier.

12.3.4   Features of half-wave pulsating and self-rectifying circuits. Chapter 13 includes details of the common controls and measuring devices in an X-ray generator; most of these are independent of the type of high-voltage rectifier circuit used. However, this is not true of the measurement of **tube current** (mA); we shall therefore describe this measurement in relation to each type of rectifier circuit.

The average X-ray exposure rate is proportional to the *average* tube *current* (10.6.2, 10.6.5), hence this is the quantity we normally wish to measure. (In diagnostic radiology we also need to measure *charge,* viz. current × time, see sections 13.4.2, 13.5.2.) Fig. 12.1 (12.2.1) and Fig. 12.2 (12.3.1) show that the self-rectifying and the two-valve half-wave circuits are both simple circuits without branches in which the X-ray tube current flows also through the high-tension transformer and is unidirectional. Therefore a moving-coil milliammeter (6.3.2) (which reads the *average* value of a current if it is varying rapidly) inserted at any point in series with the circuit will give the required result, viz. the tube 'mA'. The point at which the meter is inserted depends on electrical convenience. It is usual to place it in series with the transformer secondary at its mid-point (Fig. 12.5); here it is at earth

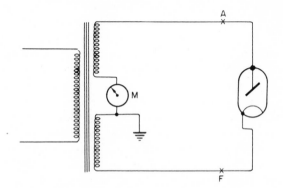

Fig. 12.5   The measurement of 'mA' in a symmetrical X-ray generator circuit.

potential and hence can be safely mounted on a control panel. If it were inserted at points A or F in Fig. 12.5, it would be at a high positive or a high negative potential relative to earth and therefore would have to be protected for safety reasons, although it would still read the correct value.

Half-wave circuits appear to have the disadvantage that they 'waste' the reverse half-cycle. However, on closer consideration it is not obvious why this *is* a disadvantage, because although there is a p.d. during this half-cycle, there is no current flowing, hence the *power* wasted is zero ($P = VI$). The real reason for the inferiority of the half-wave circuit is that *time* is being wasted; it is easy to see that for a given *average* mA (and hence a given average exposure rate) the *peak* current in the half-wave circuit would have to be twice as great as it would be in a circuit using *both* half-cycles. This, for high power generators, causes difficulties in rectifier and tube design. The next section will be devoted to a description of circuits that make use of both half-cycles of the mains supply, viz. *full-wave* rectifier circuits.

12.4 FULL-WAVE PULSATING-VOLTAGE CIRCUITS

12.4.1   The four-valve bridge circuit. Fig. 12.2a (12.3.1) shows a rectifier circuit in which two valves are used as switches. These 'connect' the transformer secondary coil to the X-ray tube during one half-cycle, in which the voltage is in the required direction. they then 'disconnect' the secondary from the tube during the opposite (bracketed) half-cycle, while the voltage is in the reverse, unwanted, direction. It is possible, however, to connect two more valves in this circuit so that the previously unused half-cycle becomes available across the tube in the correct direction; they achieve this by effectively reversing the 'bracketed' half-cycle before it reaches the tube. Fig. 12.6 is identical with Fig. 12.2 except for the addition of the two extra valves R and R'. During the *first* (forward) half-cycle of v_s, valves F and F' conduct current to the X-ray tube, valves R and R' being non-conducting. During the *second* (reverse) half-cycle of v_s, valves R and R' conduct the tube current, F and F' being non-conducting, and so on. The four valves can be thought of together as a reversing switch that reverses the transformer secondary connexions to the X-ray tube every half-cycle. As a result, during *every* half-cycle current flows in the *correct* direction through the tube, as shown.

The circuit of Fig. 12.6 is called the **four-valve bridge circuit**, and is in very

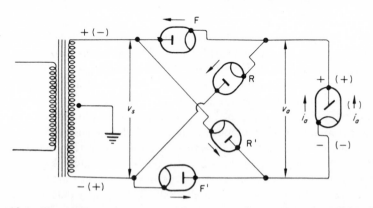

Fig. 12.6   The addition of two extra valves to the half-wave circuit of Fig. 12.2a to make a four-valve, full-wave circuit.

widespread use in large diagnostic X-ray generators. However, it is usually drawn in another form; Fig. 12.7a is exactly the same circuit as Fig. 12.6 but the diagram is rearranged and is in the form which should be memorized. The direction of electron flow during each half-cycle is shown, the dashed arrows corresponding to the bracketed half-cycles of voltage. Fig. 12.7b shows the waveforms in this circuit; note how the negative half-cycle is effectively reversed or 'inverted', and how every half-cycle produces a flow of current through the X-ray tube.

In Fig. 12.7a the method of measuring the tube current is shown as a milliammeter in series with the high-tension transformer secondary at its earthed centre point. Superficially this seems identical to the method used for

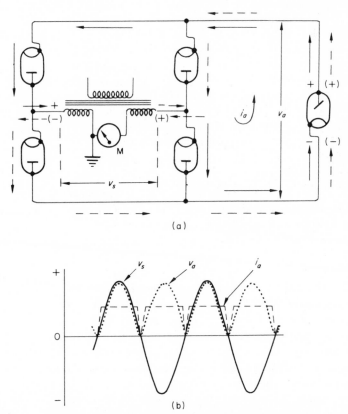

(a)

(b)

Fig. 12.7   The four-valve bridge circuit, with waveforms.

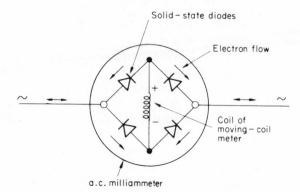

Fig. 12.8   A moving-coil (d.c.) milliammeter used in conjunction with a solid-state bridge rectifier to form an a.c. milliammeter.

the half-wave circuit (Fig. 12.5, 12.3.4). However, there is one important difference; although the *tube* current in both circuits is unidirectional, the *transformer secondary* current in the full-wave circuit is alternating. This is because both half-cycles are now in use, and because at the centre point the second half-cycle has not yet been reversed — it is reversed only through the tube. Hence the meter M (Fig. 12.7a) must be one that is capable of measuring *alternating* current. A moving-iron meter (6.3.1) *could* be used, but the many advantages of the moving-coil meter (6.3.2) can be utilized by connecting it to the circuit via a miniature 'bridge' rectifier that makes use of solid-state diodes (11.3.2). The arrangement is shown in Fig. 12.8, the 'bridge' circuit being similar to that in Fig. 12.7; as shown, the group of four rectifiers is situated inside the case of the meter. (The rectifiers are illustrated in the diagram by symbols that are used to represent rectifiers of any type; the 'arrow head' in the symbol shows the direction of *conventional* current flow.)

12.4.2   The disadvantages of all pulsating-voltage circuits. In section 10.6.1 it was explained that the quality of an X-ray beam depends, among other things, on the type of rectification and hence on the waveform of the voltage applied to the tube. A constant voltage applied to the tube will produce an X-ray spectrum whose maximum photon energy (short wavelength limit) is a function of that voltage, and whose shape depends on the total filtration in the beam. Such a beam has a certain 'penetrating power' in a given absorber. If the voltage is *decreased*, the maximum photon energy decreases, and the penetrating power *decreases*. The beam is said to be 'softer'.

Now if the voltage applied to the X-ray tube is not constant, but varying, e.g. the pulsating waveform of Fig. 12.2b (12.3.1), the maximum photon energy will be determined by the *peak* voltage, but the spectrum will contain a much greater proportion of soft radiation than did the constant voltage spectrum (10.6.1). These soft components result from those parts of the voltage half-cycle that have values below the peak, and are a considerable proportion of the total. Unfortunately, the soft components are undesirable both in radiotherapy and in radiodiagnosis, because they are absorbed uselessly in the superficial layers of the patient, perhaps even causing undesirable skin reactions. They can, of course, be filtered out (14.5); however, electrical energy has been expended in their production, and they are accompanied by undesirable amounts of heat. The most efficient method available at present for producing an X-ray beam for medical purposes is to apply a constant voltage across an X-ray tube. Methods of producing a constant voltage, commonly called a **constant potential**, will be explained in the next section.

12.5 CONSTANT POTENTIAL CIRCUITS

12.5.1 The six-valve three-phase circuit is not a true constant potential circuit but its performance is so close to the ideal that we shall discuss it under this heading. It is a circuit that makes use of the particular form in which alternating current is supplied to large buildings, viz. a form known as 'three-phase'. The word *phase* was mentioned in section 8.1.4, there referring to the *timing* of an alternating waveform. In this context it has the same meaning: a **three-phase supply** is one in which three interrelated alternating e.m.f.s are available, each 120 degrees out of phase with the other two. The details of three-phase circuits are complex and need not be explored here. The operation of a six-valve three-phase rectifier circuit can most easily be understood by development from the bridge circuit, which may be called a four-valve two-phase rectifier circuit.

Figure 12.9a shows the circuit of Fig. 12.6 (12.4.1) (i.e. one form of the four-valve circuit) redrawn in a somewhat different form, with arrows replacing the rectifiers for clarity. The arrows show the permitted direction of electron flow. The centre point of the transformer secondary is earthed and is therefore considered as being fixed at zero potential; the relative phases of the voltages at the two ends of the secondary are marked $0°$ and $180°$. The transformer primary has been omitted to simplify the diagram.

Figure 12.9b shows a six-valve three-phase generator. The transformer

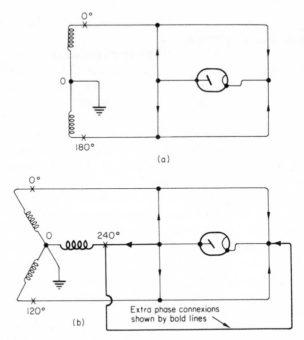

Fig. 12.9 The principles of: (a) a four-valve two-phase circuit, and (b) a six-valve three-phase circuit.

secondary now consists of three coils, the junction again at a fixed potential and the three ends supplying voltages which now have relative phases of 0°, 120° and 240°. The three coils are drawn at a spacing of 120° to emphasize this fact but this in no way represents the shape of the transformer itself. From the end of each coil, as in Fig. 12.9a, two rectifiers are connected, one to the target, one to the filament of the X-ray tube. The principle of operation of the three-phase circuit is identical with that of the two-phase, except that because of the way in which the three phases overlap in time, the total target-filament voltage varies very little from the ideal constant potential. The precise reason for this behaviour may be found in more advanced textbooks. The three-phase rectifier circuit is used in the largest and most efficient diagnostic X-ray generators.

12.5.2   Constant potential circuits with capacitance provide another means of obtaining effectively constant voltage for the X-ray tube. These circuits need not be studied in detail now, but their principle will be described, for two reasons. First, the circuits are used exclusively for 'deep' therapy generators operating at 250 kV (to be studied in Part II of the course) and second, the principle operates by chance under certain conditions in all diagnostic X-ray generators with high voltage cables.

Constant potential circuits may be regarded as being developed logically from pulsating voltage circuits, in a manner which may be clear from a water analogy. If it is desired to supply water from a well to a house, a pump may be installed in the well that will deliver water in 'pulses' of pressure. This is the analogue of the pulsating voltage generator. However, it would be most inconvenient to have a pulsating pressure from the taps in the house; instead, the pulsations must be made to provide a constant pressure (potential). This is done using the principle of a storage tank, usually situated in the roof space. The tank is filled by pulses of water from the pump; by virtue of its height (above the taps) the water then possesses potential energy (1.3.5) and a considerable flow of water (current) will result if a tap is turned on. Moreover, because the difference in levels (potential difference) between the tank and the tap is constant, the flow also will be constant. We have succeeded in producing a 'constant potential' supply of water.

The same result may be achieved in the electrical circuit by using a device that will act as a 'storage tank' for electric charge; such a device is the capacitor (3.4.2), which possesses the property of capacitance (3.4.1), and hence the ability to store charge for relatively short periods. A simple half-wave constant potential circuit is shown in Fig. 12.10a for illustration. The circuit may be considered in two separate sections: the charging section, consisting of the transformer, rectifier and capacitor (shown in thin lines), and the discharging section, consisting of the capacitor and the X-ray tube (shown in thick lines). Note that the capacitor is common to both sections.

Suppose the circuit is switched on at the instant marked 0 (Fig. 12.10b). The capacitor is uncharged ($v_c = 0$), hence as the potential of A (v_s) rises positively, the rectifier conducts and the capacitor *charges*. The potential of B (v_c) thus rises to nearly the peak value of v_s and the capacitor is fully charged. Now suppose that the X-ray tube is not conducting. As v_s begins to fall (i.e. after its positive peak), the anode of the rectifier becomes negative with respect to its filament and the rectifier ceases to conduct. Thus the capacitor cannot *discharge* through the rectifier; if the tube is not conducting, i.e. there is no discharge current, the potential of B will remain constant (Fig. 12.10b).

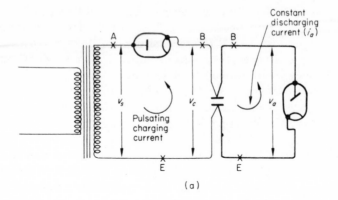

(a)

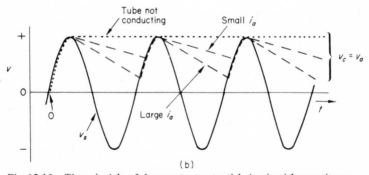

(b)

Fig. 12.10   The principle of the constant-potential circuit with capacitance.

Although we have achieved a constant potential, the condition is not very useful because the X-ray tube is not working. If now the tube filament current (and hence the filament temperature) is adjusted so that the tube can draw a small current (I_a), the polarity of the charged capacitor is such that it is in the correct direction to discharge, at a constant rate, through the tube. This causes the potential of B (v_c) to fall slowly; when the next forward half-cycle of v_s has reached the point where the potential of A rises above that of B, the rectifier conducts and the capacitor recharges nearly to the peak voltage. This process continues so long as the circuit is supplied with power, the capacitor charging in pulses from the transformer via the rectifier and discharging at a constant rate through the X-ray tube.

The tube potential is thus not strictly constant, but varies by a small amount, called the **ripple voltage**, which depends on several factors. For example, if the tube current I_a is increased (by increasing its filament temperature), the capacitor will discharge more rapidly and the ripple voltage will be larger (Fig. 12.10b). The amount of ripple, that is the extent by which the generator output falls short of the ideal constant potential, decreases

(i) as the tube current (I_a) decreases,
(ii) as the value of the capacitance increases,
(iii) as the charging frequency increases.

These factors are all taken into consideration in the design of generators for radiotherapy. Let us now see how they affect the operation of the average diagnostic X-ray generator, for example a four-valve bridge circuit.

Although these generators have no capacitors as such included in their circuits, they do have high-voltage cables that *act* as capacitors. Their inner conductor and outer metal braid (12.3.3) act as the plates of the capacitor, and the rubber insulation acts as the dielectric. The capacitance thus created is connected between the high-tension lead (either target or filament) and earth (Fig. 12.4b, 12.3.3). The value of the capacitance increases with increasing length of cable, and under average conditions is such that it has a considerable 'smoothing effect' on the pulsating voltage for low tube currents (low values of mA) such as are used for television fluoroscopy. For cables of reasonable length and for currents less than 1 mA the output can approach constant potential. However, in the traditional technique for fluoroscopy, up to 5 mA, the ripple becomes greater, and for radiography at currents of 20 mA and more the circuit behaves as a pulsating voltage generator. This type of behaviour explains the excellent results often obtained in television fluoroscopy with generators having long cables.

12.6   THE MEASUREMENT OF HIGH VOLTAGE

12.6.1   Direct methods. The value of kV used in an X-ray set, particularly in diagnostic radiology, is not normally measured directly during routine use. However, direct measurements of kV are necessary for calibration and research purposes; there are two commonly used methods, employing (i) the spark gap and (ii) the potential divider.

12.6.2   The spark gap method of measuring kV makes use of two large polished brass conducting spheres (3.3.2), mounted on insulating stands so that their distance apart can be varied and measured. The potential difference to be measured is applied between the two spheres (with a protective resistor included in the circuit); the spheres are slowly brought closer together until a spark passes. The maximum separation of the spheres at which a spark will pass is a fairly accurate measure of the peak kV (kVp). In practice, many precautions must be observed and corrections made to ensure the highest accuracy. The spark gap method, though simple and useful, has the disadvantage that it measures *only* the *peak* of the kV waveform. It is being superseded by the potential divider method.

12.6.3 The potential divider method of measuring kV is fundamentally a development of the simple voltmeter (4.3.2). Formerly it was little used because of the difficulty of manufacturing satisfactory resistors of the very high values required. However, modern technology has enabled this to be done, and the potential-divider kV-meter is having wider application.

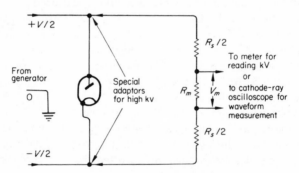

Fig. 12.11   The potential-divider method of 'kV' and waveform measurement.

Figure 12.11 shows the principle applied to a symmetrical X-ray generator circuit. A series resistor of total value R_s is split into two halves and connected in series with a low-value resistor R_m. This ensures that R_m is near earth potential. Because of the potential divider action (Fig. 4.4b, 4.3.1), the potential difference across R_m is a replica of that between target and filament but reduced in magnitude by the ratio $R_m/(R_s + R_m)$. For example, if $R_s = 1\,000\,\text{M}\Omega$ and $R_m = 1\,\text{M}\Omega$ then V_m (across R_m) will be approximately one-thousandth of V (the total kV); the p.d. V_m can then be measured in any desired way. Two types of measurement are very useful:

(i) R_m may be connected to a special type of voltmeter that reads *peak* voltage, or

(ii) R_m may be connected to an instrument called a **cathode-ray oscilloscope**. This is a device for rapidly and automatically drawing a voltage waveform on a screen (resembling a television screen). The screen may also be calibrated in volts, so that not only is the *shape* of the waveform known but also its *magnitude*. Such a measurement can give the greatest possible information about the voltage behaviour of an X-ray generator and is widely used by manufacturers for design and calibration.

13 X-Ray control and indicating equipment

13.1 INTRODUCTION

The essential parts of an X-ray generator are the X-ray tube (Chapters 10 and 11) and a high-voltage transformer and rectifier circuit (Chapter 12). So that the generator may be used effectively, control and indicating circuits must be added. These will be described in the present chapter.

The circuit of the complete X-ray generator, particularly for radio-diagnosis, is complex, and it would be quite impracticable to describe it in detail here. However, the outline circuit given in Fig. 13.3 (see diagram at the end of the chapter), though a much simplified version, demonstrates all the important features. We have given the complete circuit as a single unit rather than illustrating the various functional parts separately so that the interrelationships of one part to another will be clear from the outset.

13.2 MAINS VOLTAGE CIRCUITS

13.2.1 Mains cables, switches and fuses. In general, we shall discuss the circuit of Fig. 13.3 from left to right. At the extreme left-hand side, the live and neutral (12.3.2) connexions to the mains supply are shown. These may be of two kinds: either permanent wiring to a **junction box** in the case of a large fixed X-ray set, or a flexible cable which may be plugged into a wall socket for a smaller mobile set. In both cases, it is very important that the mains cables are of adequate thickness and therefore of low enough resistance

184

to carry without excessive **voltage drop** (5.3.2) the large currents that may flow.

In Table 11.1 (11.3.4), for example, it is shown that the power fed to the tube during a large radiographic exposure might be as much as 50 kW. Assuming the generator circuit to be 100% efficient (which in practice it is not), the power drawn from the mains supply at, say, 250 V (r.m.s.) will also be 50 kW. Hence the current in the mains cables will be $I = P/V =$ 50 000/250 = 200 A! If we specify that the voltage drop during the exposure shall not be greater than, say, 10 V, the resistance of the mains cables must not be greater than $R = V/I = 10/200 = 0.05$ Ω. This is quite a low resistance, particularly if it has to be attained in long cables (4.2.3). The problem is especially difficult for mobile generators (though they do not draw as much as 200 A) for three reasons: (i) they are connected by flexible cable whose thickness must be limited, (ii) the plug and socket must make excellent contact of low resistance, and (iii) it must be certain that any of the wall sockets to which the set is connected are wired with cables of adequate thickness. All X-ray generators are calibrated and adjusted by the manufacturers on the assumption that the mains voltage is constant; excessive mains voltage drop will result in the kVp during the exposure being less than intended, causing unsatisfactory radiographs.

The mains supply passes to the generator circuit via a **double-pole** mains switch. This, as shown in Fig. 13.3, consists of two switches mechanically linked, i.e. operated by the same knob, so that they are both 'on' or both 'off' together. This switch serves the dual purpose of switching the whole set off either at the end of the day or when it is necessary to isolate the circuits from the mains supply for safety during maintenance.

After the main switch come the fuses whose conventional symbol looks like an 'infinity' sign (∞); these serve to protect both the generator and the mains wiring in the event of a serious fault developing (5.2.2).

13.2.2 Mains (line) voltage compensation. The mains voltage, or as it is sometimes called, the **line voltage**, has a nominal r.m.s. value in the U.K. of 240 V. However, the r.m.s. value is not normally constant but fluctuates, sometimes exceeding the nominal value but more often falling below it. This fluctuation will usually seriously affect the functioning of an X-ray set and steps must be taken to reduce its effects.

Mains fluctuations are of two general types: (a) large, slow fluctuations which might be as much as 10 or 15 V and might extend over several hours;

these are related to the well known 'peak periods' which result from industrial working hours, domestic cooking times, cold weather spells, etc.; and (b) small, fast fluctuations of the order of a volt or less occurring over periods of seconds or less; these result from the switching on or off of other X-ray sets, lifts, etc.

The large, slow fluctuations would affect the operation of the whole set and must be dealt with at the start. This is done by taking the live (L) mains lead to a selector switch, called the **line voltage compensator** (Fig. 13.3, page 192), which selects one of a number of tappings on a large autotransformer (8.3.3). As the mains voltage varies slowly, the line voltage compensator can be so adjusted that the potential difference across the whole autotransformer (or a fixed part of it) remains approximately constant. Thus a connexion taken from the upper end of the autotransformer (in the example of Fig. 13.3) can serve as a 'live' mains supply lead with a fairly constant voltage; this we may call the 'live compensated' lead (LC). The line voltage compensator is periodically adjusted by the radiographer by reference to an a.c. voltmeter, the **line voltmeter**, connected between LC and neutral N, the pointer of which must be kept between two lines marking the permissible limits of voltage variation.

The small, fast fluctuations affect only a small part of the circuit and will be discussed later (13.5.1).

13.3   X-RAY TUBE VOLTAGE (kV)

13.3.1   The resistance method of kV control. It would at first sight seem logical to control the voltage applied to the X-ray tube (the kV) in the high voltage circuit itself. However, controls operating at high values of kV are difficult to design and are often unreliable (because of insulation problems); in practice it is simplest to control kV by varying the input voltage to the primary of the high-voltage (high-tension) transformer.

Several methods are available for this purpose, of which one is shown in Fig. 13.1. A rheostat R is put in series with the transformer primary. The primary current I_p produces a voltage drop V_r across R, such that $V_r = I_p R$. Then the input voltage to the transformer V_p is equal to the initial voltage V_1 less the voltage drop V_r, or

$$V_p = V_1 - I_p R.$$

Eq. 13.1

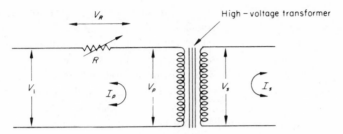

Fig. 13.1 A simple but unsatisfactory method of 'kV' control.

As V_1 may be assumed constant, the primary voltage V_p, and hence the kV on the tube, may be controlled by varying R.

This simple arrangement has two major disadvantages. First, because a current is flowing through a resistance, there will be power equal to I_p^2R (5.2.1) converted into heat and therefore wasted. Second, and more important, the product I_pR, and hence V_p itself, depends on the primary current I_p. But this in turn depends on the secondary current I_s (8.2.2) and hence on the tube mA. Thus if R is set to give a particular value of kV, and the mA is subsequently changed, the kV will change by a very large amount. The mA and kV would thus be seriously interdependent (contrast 11.2.2). This type of behaviour is reminiscent of the simple cell or other generator described in section 5.3.2; it results from the fact that the inclusion of R in the circuit enormously increases the effective internal resistance of the high-voltage generator. Clearly a method of varying the voltage is required which will be almost independent of the current flowing, viz. which will have the smallest possible internal resistance. Such a method is described in the next section.

13.3.2 Pre-reading kV control and indication. The best method of providing a variable input to the high-voltage transformer primary is to use a separate variable-ratio transformer. As the circuit (Fig. 13.3, page 192) already contains an autotransformer, it is convenient to provide this with extra tappings connected to another selector switch called the **pre-reading kV control.** By adjusting the switch, the autotransformer ratio is altered, and its output voltage is therefore varied. Sometimes an a.c. voltmeter is connected between the tapping point and the neutral main N; this voltmeter can be calibrated directly in kV (although of course it is actually measuring transformer primary voltage) and is therefore called a **pre-reading kV meter.**

187

The methods described in section 12.6 for measuring kV are of course not applicable to clinical work, particularly radiography, because it is necessary to set the kV on the generator *before* the exposure is made whereas the direct methods of measuring kV operate only *during* the exposure.

In the circuit in Fig. 13.3, and in most modern generators, a separate pre-reading voltmeter is not required. This is because the action of the line voltage compensator (13.2.2) always ensures that the autotransformer is energized to a constant level (within certain limits of error); therefore the tappings on the kV selector always represent a known voltage. The tappings themselves can therefore be labelled with their corresponding kV values; this results in the familiar control knob marked directly in kV.

The autotransformer type of voltage control does not suffer from the principal disadvantage of the rheostat control (13.3.1), viz. interdependence of mA and kV, because the autotransformer has a very low internal resistance (5.3.2). It therefore has very good *regulation* (8.2.3).

Figure 13.3 shows the lower end of the high-voltage transformer primary connected to a second selector switch instead of to the neutral main. The reason for this will be explained in section 13.5.3.

13.4 EXPOSURE CONTROL

13.4.1 Contactors and timers. The X-ray exposure is usually controlled by switching the high-voltage transformer primary current on and off. We have shown (13.2.1) that the mains current, and therefore the primary current, might be very large — as much as 200 A. This is too large to be switched conveniently by a simple switch, hence a **contactor** (6.2.3) is used, itself controlled by a small pair of contacts in one form or another. If manual operation is required, for example in fluoroscopy, the small contacts will take the form of a switch. If automatic exposure timing is required, the small contacts will form part of an **exposure timer**. To control exposure *time* (in seconds or minutes) the exposure timer will be a form of clock mechanism fitted with contacts (see below). To control either *mAs* or *exposure in roentgens* (10.5.3) directly, some form of autotimer (13.4.2) may be used. This group of components is shown in 'block diagram' form between the kV control and the high-voltage transformer in Fig. 13.3 (page 192).

In practice it is more efficient to have two contactors, with their contacts in series, one to switch the exposure on, the other to switch it off. Moreover, the instant of switching is usually timed to coincide approximately with the zero of the alternating current cycle so that the contacts have to handle the minimum possible current. In 250 kV radiotherapy generators, it is undesirable to switch the full kV on to the tube instantaneously. Instead, the kV is increased gradually to its maximum by a clock-controlled series of contactors; then the radiation exposure is started by opening a lead shutter in the beam. The exposure is terminated by switching off the current in the usual way.

The exposure timer for controlling *time* of exposure consists of a **synchronous** electric motor operating a pair of contacts. The synchronous motor is of the kind used in electric clocks, which keeps in step with the cycles of alternating current and which therefore maintains a constant speed. The motor revolves continuously, controlling the opening and closing of the contacts via a pair of friction disks, one of which must be set each time to the desired exposure. Mechanical details of synchronous clock timers may be obtained from books on X-ray engineering.

Mechanical timers of this kind are satisfactory except for very short exposures or for a rapid succession of exposures such as in serial radiography. They may then be replaced by an **electronic timer**, which is a combination of a resistance-capacitance charging circuit (4.3.3) with a cold-cathode gas-filled diode or similar tube (11.2.3). Fig. 13.2a shows the principle of an electronic timer. Initially Sw is closed and current flows from V_B through R and through Sw. The p.d. across C (V_c) remains at zero. At the start of the exposure Sw is opened and the current which formerly flowed through it is diverted to the capacitor C which commences to charge. V_c increases exponentially (4.3.3); when it reaches the value at which the trigger tube 'fires', a current passes through the relay Ry whose contacts open thus terminating the exposure. The period between Sw opening and the contacts of Ry opening is equal to the exposure *time;* it is determined by the values of R and C (4.3.3). The electronic timer can produce exposure times down to one thousandth of a second. The trigger device in Fig. 13.2a is shown in 'block' form because, although in principle a cold-cathode *diode* (11.2.3) could be used, in practice a better performance is obtained from more complex tubes whose details we cannot discuss here (cold-cathode triodes, thyratrons).

13.4.2 Autotimers. The basic timer principle illustrated in Fig. 13.2a (13.4.1) may be used in a variety of ways to control the X-ray exposure in terms of factors other than time itself. This may be done in three principal ways.

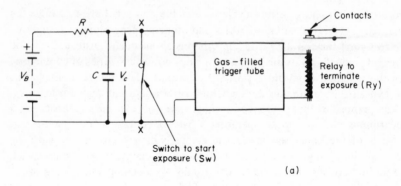

(a)

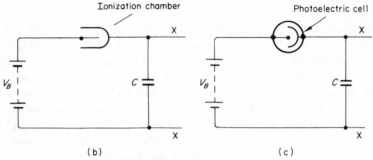

(b)   (c)

Fig. 13.2   Various types of electronic timer.

(a) The total exposure = exposure rate × time (10.5.3, 10.6.5), and the exposure rate is proportional to mA; therefore total exposure must be proportional to mA × time or **mAs**. Exposure time may be regulated to give a particular value of mAs by allowing the capacitor C in Fig. 13.2a to be charged by a current proportional to the X-ray tube current or mA, instead of by V_B and R. This current is usually derived from the mid-point of the high-voltage transformer secondary, with rectification if necessary (12.3.4). The associated charge is stored in the capacitor which is so chosen that at the required value of mAs the triggering voltage is reached; the exposure is then terminated. Different values of mAs may be selected by choosing different values of capacitor by means of a switch.

(b) Any required value of radiation exposure, measured at any location in the beam, can in principle be selected by placing an ionization chamber (15.2.3, 15.2.4) at the desired location (e.g. at the exit aperture of the X-ray tube housing) and allowing the ionization current to charge the capacitor C (Fig. 13.2b). In practice, the simple arrangement shown in the diagram may not be adequate because the ionization current in certain conditions may be too small to operate the cold-cathode tube. Then, a different approach is adopted using an **electrometer valve**, but the principle is the same. (An electrometer valve is a development of the diode which enables very small values of charge, such as can be measured on an electrometer (3.2.1), to give an indication on a moving-coil meter.) One application of this method is the exposure timer used in radiotherapy treatments. In this, the ionization chamber is situated in the exit aperture of the X-ray tube housing and results in a constant field exposure being delivered to the patient.

(c) In the diagnostic situation the ionization chamber of Fig. 13.2b may be positioned near a film cassette so that the X-ray exposure time results in a constant radiation exposure to the cassette. This can be so regardless of kV, mA, patient thickness, etc. and in principle results in a constant photographic exposure and therefore constant density in the film. Another way of achieving this result is embodied in the **phototimer**. A fluorescent screen in a light-tight box, situated behind the film cassette (which must not contain a lead screen, 14.6.3 (ii)) is irradiated by the X rays that pass through the film and its intensifying screens. During the exposure, the fluorescent screen emits a total amount of light which is proportional to the cassette exposure. The light falls on the cathode of a device known as a **photoelectric cell** (Fig. 13.2c). The cathode has the property of emitting electrons (**photoelectrons**), i.e. an electric charge, whose magnitude is proportional to the total amount of light and hence to the cassette exposure. These electrons are collected by the anode of the photoelectric cell, resulting in the capacitor C being charged as before; the phototimer therefore terminates the exposure period when a constant exposure has been delivered to the cassette. The phototimer is claimed to have the advantage over the ionization chamber timer that it measures light directly, viz. the same type of energy that exposes the film between its intensifying screens (10.1 (ii)), and therefore changes of X-ray quality should not affect its operation.

13.5 X-RAY TUBE CURRENT (mA)

13.5.1 mA control and stabilization. Because the X-ray tube acts as a diode operated in its saturated or temperature-limited condition (11.2.2), change of kV in theory has no effect on the value of mA. In practice, the X-ray tube is not perfectly saturated, and therefore departs somewhat from the above ideal behaviour. This requires the use of compensating circuits which will not be discussed here.

However, if we wish to vary the mA at will, we must do so by changing the filament current, resulting in a change in filament temperature. This is achieved as shown in Fig. 13.3 (page 192). The tube filament current is supplied by a voltage step-down transformer which is of the highly insulated type referred to in section 8.3.2 (iii). The filament current is altered by varying the value of resistance in the filament transformer primary circuit. The selector switch (mA CONTROL) in Fig. 13.3 enables the operator to select either (a) one of a number of **pre-set** (i.e. previously adjusted) resistors, corresponding to distinct radiographic currents, e.g. 50, 100, 200, 500 mA, or (b) a rheostat which enables fluoroscopic current to be varied continuously. In radiotherapy the possibility of continuous variation of current is usually provided but a single value, e.g. 15 mA at 250 kV, is commonly used.

While the main switch of the set is on, but the X rays are switched off, the filament of the X-ray tube is kept hot in readiness for an exposure. However, the fraction of the total time devoted to exposures is relatively small, and if the filament were kept at its full temperature for the *whole* of the time, excessive evaporation of tungsten would take place and the life of the filament would be greatly reduced. Instead, the filament is kept at a lower 'stand-by' temperature by including extra resistance in series with the mA control; just before the exposure, in the condition known as 'prepare', the filament temperature is raised to its correct value. This technique is called **filament boost.**

One end of the filament transformer primary is connected to the neutral main. The other end would logically be connected to the live compensated main LC. However, this mains lead still carries the small fast mains voltage variations (13.2.2); these, although small, are large enough to cause serious instability in the X-ray tube current (mA). This is because the saturated electron emission from the filament is critically dependent on the filament temperature and therefore on the filament current (11.1.1). To reduce this effect, a unit called a **voltage stabilizer** is interposed between LC and the mA control switch (Fig. 13.3). The operation of this device is too complex to be discussed here. It reduces the mains voltage fluctuations by a factor of ten or fifteen; the variations at the point LS ('line stabilized') are thus small enough to be unimportant.

The electron emission from the rectifying valves is not at all critical because it is made sufficiently copious to keep the valve resistance low in the forward direction. This is done by heating the filaments to a sufficiently high temperature (11.3.1 (ii)). Hence it is unnecessary to operate the valve filament transformers from the stabilized supply; Fig. 13.3 shows how they are connected instead to LC, the live compensated main.

13.5.2   mA and mAs indication. The common method of measuring average tube current (mA) is described in sections 12.3.4 and 12.4.1, viz. a moving-coil meter used with an additional solid-state rectifier circuit if necessary. However, in a radiographic exposure it is not so important to know the exact values of mA and exposure time separately; the important factor for film exposure is their product, mAs, which is identical with electric *charge* (4.1.4).

The moving-coil meter can be readily adapted to measure charge or mAs *directly;* this is done by making its moving system (viz. the coil, pointer, etc.) of relatively large mass and by reducing the restoring force of the hair springs (6.3.2) almost to zero. Then if a pulse of current of short time duration, such as a radiographic exposure, is passed through the coil, the resulting deflexion will be proportional, not to the current, but to the *total charge* in the pulse. The scale is then calibrated in mAs. The pointer remains at its final reading after the pulse until it is reset to zero, usually by a reverse current passed through the coil.

13.5.3   Generator regulation and mA compensation. High-voltage generators for X-ray tubes behave in the same way as simple cells (5.3.2), mains supplies (13.2.1) and many other forms of generator in that their output voltage tends to fall as a greater current is drawn from them. This phenomenon results from the internal resistance of the generator (5.3.2); in this case the internal resistance is made up of contributions from all the components that make up the generator. As with transformers (8.2.3), the constancy of the generator voltage output with change of mA is called its **regulation**; a generator with *good* regulation maintains its voltage output for larger changes of mA than one with *bad* regulation.

However, every generator has *some* internal resistance; the resulting change of kV with mA is not acceptable in radiology because it causes a change in radiological contrast. The generator must therefore be compensated. This is done by means of the **mA compensation** circuit shown in Fig. 13.3 (page 192). As an *increase* of mA would normally result in a *decrease* of kV, this

193

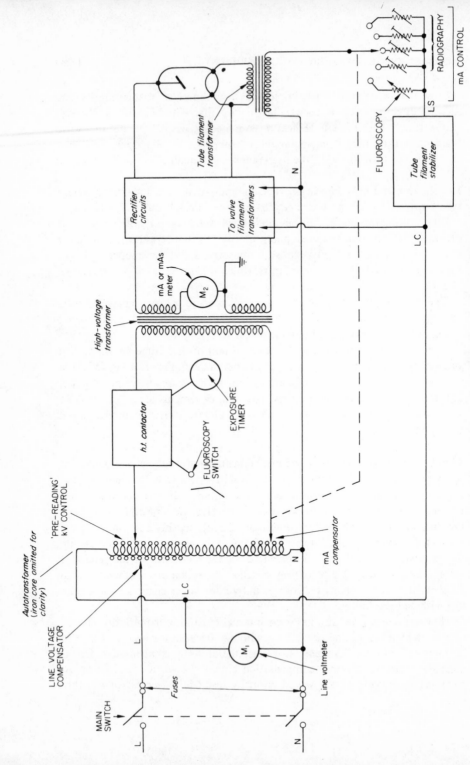

Fig. 13.3 An outline of the control and indicating circuits of an X-ray generator.

may be compensated for prior to the exposure by slightly increasing the voltage input to the high-voltage transformer primary. This is achieved by mechanically linking the selector switch marked 'mA compensator' to the mA control switch. The positions of the tappings on the autotransformer are selected by the manufacturers during the generator design to give just the correct amount of compensation.

14 Interaction of X rays and gamma rays with matter

14.1 THE SEQUENCE OF EVENTS

14.1.1 The absorption of energy. In order to understand the absorption of radiation, which forms the basis of radiology and of radiotherapy, it is necessary to consider the ways in which radiation interacts with any object through which it is passing.

When a beam of X rays or gamma rays passes through matter, the interaction results in a reduction in the intensity (i.e. the energy per unit area per unit time) of the beam. Some of the energy is *absorbed* by the medium and some is *scattered* out of the beam (14.2.2). The energy which is *absorbed* by the medium causes the changes which we observe in the medium, e.g. chemical and biological changes. The sequence of events which results in these changes is summarized in Fig. 14.1. The energy is absorbed by various processes (14.3) and is converted initially into kinetic energy of electrons in the medium. The electrons then move through the medium producing ionization and excitation of atoms and molecules (2.4) resulting in turn in chemical changes.

14.1.2 Biological changes. If the radiation is passing through biological material, the ionization and excitation may occur in biologically important molecules, thus damaging or killing cells by **direct action**. Alternatively, cells may be damaged or killed as a consequence of chemical changes produced in the medium *surrounding them* which is largely water; this is known as

196

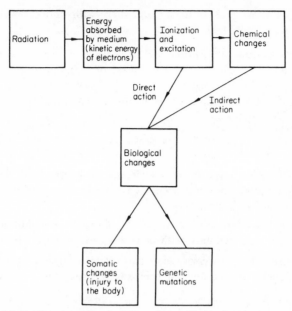

Fig. 14.1 (14.1.1) The sequence of events when radiation is absorbed by a biological medium.

indirect action. The resulting biological changes are called **somatic** if they are evident in an individual's body during his life time, e.g. a radiation burn, and **genetic** if they are mutations of the genes (or certain other changes) in the germ cells. (These are the cells which govern the inherited characteristics of future generations.)

14.2 THE TRANSMISSION OF A HOMOGENEOUS BEAM THROUGH A MEDIUM

14.2.1 Homogeneous and heterogeneous beams. A beam of X rays or gamma rays may be regarded as a stream of photons, each photon having energy of a certain value (9.2.2). In a **homogeneous (monochromatic** or **monoenergetic)** beam (10.5.1) all the photons have the *same energy*, as, for example, can be the case for gamma radiation (16.3.3). In a **heterogeneous** beam, however, the photons are *not* all of the same energy; their energies cover a range of values as, for example, in the continuous spectrum of X radiation (10.5.1).

We shall deal first with the attenuation of a homogeneous beam of

197

radiation, because it is simpler, and then with the attenuation of a heterogeneous beam (14.4).

14.2.2 Attenuation: absorption and scattering. These three terms have definite meanings when applied to a beam of radiation passing through a medium.

DEFINITION   **Attenuation is the reduction of the intensity of a beam as it passes through a medium.**

Attenuation is due to absorption or to scattering or to a combination of both (14.3).

DEFINITION   **Absorption is the transference of energy from the radiation to the medium; after transfer, the energy is present in the medium initially as the kinetic energy of electrons.**

DEFINITION   **Scattering is a change in the direction of a photon (due to an interaction with the medium), with or without loss of energy by the photon.**

In some attenuation processes, secondary electromagnetic radiation or re-emission of energy by the medium occurs; this is described in sections 14.3.3 and 14.3.5.

14.2.3 The exponential law. As a parallel* beam of *homogeneous* radiation passes through a uniform medium, *equal thicknesses* of the medium attenuate the beam by a *constant fraction*. If, for example, a 10 mm thickness of a medium attenuates a beam by a factor of one half, then the remaining half of the radiation is transmitted *through* that thickness. The next 10 mm of medium will attenuate the beam by a further factor of one half, so that one-quarter of the initial intensity remains after passing through a total of 20 mm. A further 10 mm of medium will attenuate the radiation by one half again, so that one-eighth of the initial intensity is transmitted through 30 mm, and so on.

This type of relationship is known mathematically as an **exponential law**. It is illustrated graphically in Fig. 14.2 where the intensity of the radiation transmitted, expressed as a percentage of the initial intensity, is plotted against the thickness of the medium traversed. This **transmission curve** can be represented by the equation,

$$I = I_0 e^{-\mu t},$$ Eq. 14.1

*For a *parallel* beam, the reduction in the intensity is due solely to the attenuation by the medium. For a *diverging* beam, an additional reduction in the intensity occurs due to the inverse square law (9.1.4).

where I is the intensity transmitted through thickness t,
 I_0 is the initial intensity of the beam, i.e. when $t = 0$,
 e is a mathematical constant, and
 μ is the **total linear attenuation coefficient** of the medium for the energy of the particular radiation (14.2.4).
Note that Eq. 14.1 is analogous to Eq. 16.1 (16.5.1), which represents the exponential decay with time of the activity of a sample of a radionuclide, and to Eq. 4.7 (4.3.3), which represents the exponential discharge with time of capacitance through resistance.

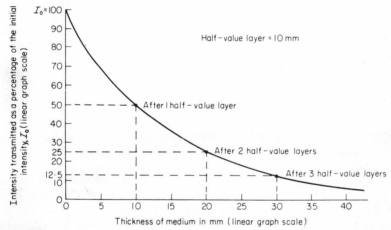

Fig. 14.2 A transmission curve, plotted on linear graph scales, for homogeneous radiation passing through a medium.

In Fig. 14.2, the transmission curve is plotted with *linear scales* for both axes of the graph. If the curve is replotted using a *logarithmic scale* (i.e. one in which equal *distances* along the axis represent equal *ratios* of the quantity) for the intensity axis and a *linear scale* for the thickness axis, a straight line graph is obtained (Fig. 14.3). This is a mathematical property of the exponential law. The slope of the straight line is numerically equal to μ.

14.2.4 Attenuation coefficients. The effectiveness of a medium as an attenuator or 'absorber' of the radiation passing through it is measured in terms of an **attenuation coefficient** such as the total linear attenuation coefficient μ.

199

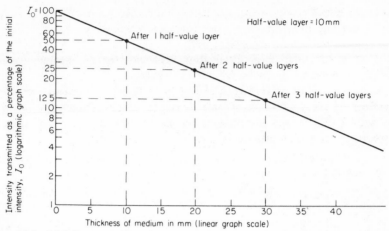

Fig. 14.3 (14.2.3) A transmission curve, plotted on log-linear graph scales, for homogeneous radiation.

DEFINITION   The total linear attenuation coefficient is the fractional reduction in the intensity of a parallel beam of radiation per unit thickness of the medium traversed.

This definition can be deduced mathematically from Eq. 14.1 (14.2.3).

The coefficient μ is called *total* because it includes *all* the processes (14.3) by which the radiation is attenuated. It is called *linear* because the attenuation per unit *thickness* of medium is being considered. The value of μ for a given medium and for a homogeneous beam of given energy can be calculated if the half-value layer is known using Eq. 14.2 (14.2.5). The units for μ are 'per unit distance', e.g. mm^{-1} or cm^{-1}, because it is defined as a fraction (a number without dimensions) per unit distance.

Among other things, the value of μ depends on the number of atoms per unit volume of the medium; therefore if the medium is compressed or changes its physical state from solid to liquid or to gas, the value of μ changes. However, the ratio μ/ρ, where ρ is the density (1.2.2) of the medium, does not change with the physical state of the medium because ρ also depends on the number of atoms per unit volume. μ/ρ is therefore more representative of the *elements present* in the medium and is known as the **total mass attenuation coefficient** of the medium.

DEFINITION   **The total mass attenuation coefficient is the fractional**

reduction in intensity of the radiation per unit mass of the medium traversed by a parallel beam of unit cross-sectional area.

The dimensions of μ/ρ are L^2M^{-1} (1.2.2) and the units in almost universal use are $cm^2 g^{-1}$. (In S.I. units, they would be $m^2 kg^{-1}$, where $1 \, m^2 kg^{-1} = 10 \, cm^2 g^{-1}$).

14.2.5   Half-value layer (or thickness). In sections 10.5.1 and 10.5.2, it was explained that the quality of a beam of radiation can be described by stating the thickness of a given material which reduces the exposure rate (15.2) of the beam to one half. This thickness is known as the **half-value layer** (h.v.l.) or **half-value thickness** (h.v.t.) of the beam in that particular material; an experimental method for measuring it is described in section 15.3.

In the transmission curves shown in Figs 14.2 and 14.3 (14.2.3), the intensity* of the homogeneous beam is reduced to one half by a thickness of 10 mm. The half-value layer of the beam of radiation is therefore 10 mm in that particular material.

For homogeneous radiation, the total linear attenuation coefficient μ can be calculated from the half-value layer using the equation:

$$\mu = \frac{0 \cdot 693}{(\text{h.v.l.})}.$$ Eq. 14.2

This equation is obtained by substituting $I = I_0/2$, and $t = $ (h.v.l.) in Eq. 14.1 (14.2.3), then rearranging the equation with μ on the left-hand side.

Note that Eq. 14.2 is analogous to Eq. 16.2 (16.5.2) which relates the radioactive decay constant of a radionuclide to its half-life.

14.3   ABSORPTION AND SCATTERING PROCESSES

14.3.1   Introduction. There are five different processes by which X rays and gamma rays may be absorbed or scattered when they pass through a medium. These are

   (i)   classical or unmodified scattering,
   (ii)   photoelectric absorption,
   (iii)   Compton or modified scattering,
   (iv)   pair production,
   (v)   photo-nuclear disintegration.

*The relationship between intensity and exposure rate is dealt with in section 10.5.3. For homogeneous radiation, one is proportional to the other and consequently the half-value thickness may be defined in terms of either intensity or exposure rate.

The amounts of absorption and scattering and the proportions of the individual processes which occur when a beam passes through a medium depend on: (a) the energy of the radiation, and (b) the atomic number of the medium, as explained below for each of the processes.

14.3.2   Classical or unmodified scattering can occur when the photons in the beam have energies which are small compared with the binding energies (2.4) of electrons in the atoms of the medium. The process therefore takes place mainly with radiation of long wavelength (i.e. low photon energy) and a medium of high atomic number.

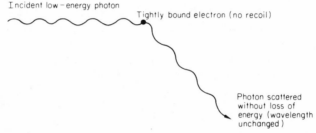

Incident low – energy photon

Tightly bound electron (no recoil)

Photon scattered without loss of energy (wavelength unchanged)

Fig. 14.4   Classical or unmodified scattering.

An incident photon collides with an electron and rebounds *without* causing it to recoil because the electron is tightly bound in a shell of an atom (Fig. 14.4). No kinetic energy is acquired by the electron because it does not recoil. Consequently the photon is scattered *without* loss of energy and therefore with *unmodified* wavelength.

In this process, therefore, scattering occurs but there is no absorption of energy by the medium. The process usually makes only a small contribution to the attenuation of a beam in radiology.

14.3.3   Photoelectric absorption can occur when an incident photon has energy equal to or greater than the binding energy of an electron in an atom of the medium. In this case the photon can ionize the atom by ejecting an electron from a shell. The incident photon gives up *all* its energy to the atom. The electron, known as the **secondary electron** or **photoelectron**, is ejected with kinetic energy equal to the energy of the incident photon minus the binding energy (the latter has to be supplied to remove the electron from the shell). The vacant site in the shell is then filled by an electron jumping inwards from another shell farther away from the nucleus. This transition is accompanied by the emission of **characteristic X radiation** in the form of a secondary photon whose energy is equal to the difference between the binding energies of the two shells involved in the transition (Fig. 14.5). The characteristic radiation is identical in nature to that generated in the target of an X-ray tube (10.3 (iii)).

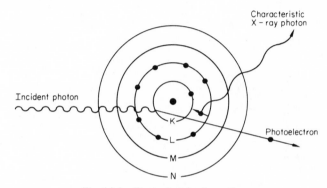

Fig. 14.5   Photoelectric absorption.

In this process, no scattering occurs because the incident photon gives up all its energy and ceases to exist. (The secondary characteristic X-ray photon is *not* the incident photon scattered.) Energy is absorbed and is present in the medium initially as the kinetic energy of the photoelectron. This then moves through the medium giving up its energy by producing ionization or excitation in other atoms.

τ* is used to represent the linear attenuation coefficient for the photoelectric process *taken alone;* hence τ/ρ represents the mass attenuation coefficient for the photoelectric process. The value of τ/ρ for a given medium varies with the photon energy in the manner shown in Fig. 14.6. For photons with energies which are high compared with the binding energy for the K-shell of atoms of the medium, the value of τ/ρ is quite small. As the photon energy decreases, τ/ρ rises to a sharp peak at a value of photon energy equal to the binding energy for the K-shell, i.e. photoelectric absorption

*τ is the Greek small letter tau.

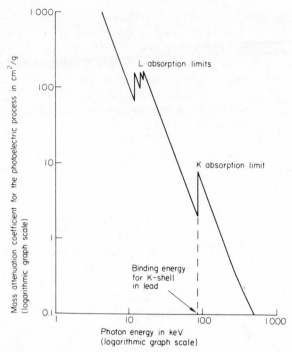

Fig. 14.6   The mass attenuation coefficient for the photoelectric absorption process in lead, plotted against photon energy on log–log graph scales.

involving electrons in the K-shell is greatest when the photon energy equals the binding energy for the K-shell. At a photon energy just less than this value, τ/ρ drops sharply because the photons have insufficient energy to eject electrons from the K-shell; τ/ρ then represents photoelectric absorption in the L-shell. As the photon energy decreases further, τ/ρ rises again until the photon energy equals the binding energy for the L-shell; another sharp discontinuity then occurs in the graph. (The discontinuity at this point may have more than one peak, such as the three peaks in the case of lead, because there can be sub-shells in the L-shell. These have slightly differing values of binding energy and each sub-shell has a corresponding peak.) The discontinuities in the graph are known as the **K absorption limit** or edge and the **L absorption limits** or edges.

If the photon energy is kept constant and the medium changes, τ/ρ is proportional to Z^4/A, where Z and A are the atomic number and the mass number respectively of the medium (2.1.4). Z/A is, however, approximately constant for all elements and therefore, for a given photon energy, τ/ρ is approximately proportional to Z^3, i.e. the mass attenuation coefficient for the photoelectric process varies greatly with the atomic number of the medium. This marked variation has great significance in radiotherapy (14.6.2) and in diagnostic radiology (14.6.3).

14.3.4   Compton or modified scattering can occur when an incident photon has much greater energy than the binding energy of an electron in an atom of the medium, so that the electron behaves as if it were free. On colliding with the electron, the photon causes it to recoil; thus part of the energy of the photon is transferred to the electron. The photon is scattered and continues with reduced energy and therefore with modified wavelength (Fig. 14.7). The recoil electron is known as the **secondary** or **Compton electron.**

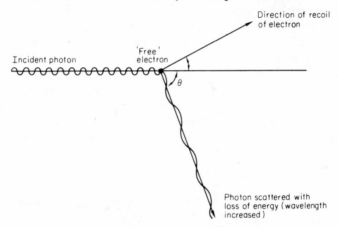

Fig. 14.7   Compton or modified scattering.

In this process, both scattering of the photon and absorption of energy take place. The absorbed energy is present in the medium initially as the kinetic energy of the secondary electron; this then moves through the medium giving up its energy as it produces ionization or excitation in other atoms.

The increase in the wavelength of the scattered photon depends on the angle θ through which it is scattered. The increase is greater the larger the angle, and is greatest when the angle is 180 degrees, i.e. a photon loses most energy when it is scattered back along its original path. The increase in the wavelength depends *only on the angle* and is *independent* of the composition of the medium and of the actual wavelength of the photon. (For example, the increase in the wavelength is always 0·0024 nanometre for radiation scattered through 90 degrees.) In consequence, when short wavelength (high energy) radiation is scattered through a given angle, the *percentage* change in the wavelength is greater than when long wavelength (low energy) radiation is scattered through the same angle. In other words, when *short* wavelength radiation is scattered by the Compton process, the change in the quality is more marked than when *long* wavelength radiation is scattered.

σ^* is used to represent the linear attenuation coefficient for the Compton process taken alone; hence σ/ρ represents the mass attenuation coefficient for the Compton process. The value of σ/ρ for a given medium decreases continuously with increasing photon energy. This is shown in Fig. 14.8 which is a graph of (i) the attenuation coefficients for each of the processes taken separately and (ii) the total attenuation coefficient (i.e. the sum of the individual coefficients (14.3.7)) plotted against photon energy for a homogeneous beam of X rays.

If the photon energy is kept constant and the medium changed, σ/ρ is proportional to Z/A. This, however, is approximately a constant for all elements and therefore, for a given photon energy, σ/ρ is almost *independent* of the composition of the medium. In other words, the mass attenuation coefficient for the Compton process is approximately the same for all media.

14.3.5 Pair production can occur when an incident photon has energy equal to or greater than 1·02 MeV. At these energies, a photon can interact with the field around the *nucleus* of an atom of the medium; this results in the *spontaneous creation* of a pair of electrons, one *negatively* charged and the other *positively* charged (see below).

In this process, mass and energy are seen to be *interchangeable* (1.1.4). Einstein showed that they are related by the equation,

$$\mathcal{E} = mc^2 \,, \qquad\qquad \text{Eq. 14.3}$$

where $\mathcal{E}$ is the energy needed to create a mass m, and c is a constant equal to the velocity of electromagnetic waves in vacuum (about 3×10^8 m s^{-1}). Conversely, $\mathcal{E}$ is the energy that will be produced if a mass m is *completely destroyed or annihilated.*

---

*σ is the Greek small letter sigma.

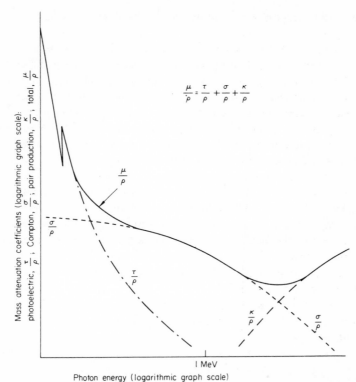

$$\frac{\mu}{\rho} = \frac{\tau}{\rho} + \frac{\sigma}{\rho} + \frac{\kappa}{\rho}$$

Fig. 14.8 (14.3.4) The mass attentuation coefficients for the individual processes, and the total mass attenuation coefficient, plotted against photon energy on log-log graph scales.

If the mass of two electrons is substituted in Einstein's equation, the equivalent energy is found to be 1·02 MeV. Hence a photon having an energy of 1·02 MeV is needed to create a pair of electrons. If the photon has more energy than this, the excess energy is converted into the kinetic energies of the two electrons created.

Of the pair of electrons created in this process, one is of the type normally encountered, e.g. in the shells of an atom (2.1.3). It is known as a **negatron** and has a negative electric charge (−1 atomic unit). The other type of electron is known as a **positron**; it has the same mass as a negatron but has a positive charge (+1 atomic unit). The positron is an unusual particle and is only encountered in the special circumstances of pair production and of radioactivity (16.3.2).

207

When a photon produces a pair of electrons by this process, both electrons move through the medium giving up their kinetic energies by producing ionization or excitation in atoms of the medium. When the negatron slows down, it is indistinguishable from the other ordinary electrons in the medium. When the positron has given up all its kinetic energy and has come to rest, it combines with a negative electron in the medium and they are *annihilated.* This process of **annihilation** is the reverse of creation; the combined masses of the positron and the negative electron are converted into 1·02 MeV of energy. This energy is radiated as *two photons,* each of energy 0·51 MeV, travelling *at 180 degrees to each other,* i.e. in opposite directions. These photons are known as **secondary annihilation radiation**.

The overall process of pair production followed by annihilation is illustrated in Fig. 14.9. In this process, no scattering occurs because the incident photon gives up *all* its energy and ceases to exist. Energy is absorbed and is present in the medium initially as the kinetic energies of the negatron and of the positron.

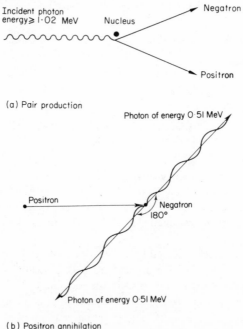

(a) Pair production

(b) Positron annihilation

Fig. 14.9   Pair production followed by positron annihilation.

κ* is used to represent the linear attenuation coefficient for the pair production process; hence κ/ρ represents the mass attenuation coefficient for the pair production process. The value of κ/ρ for a given medium is *zero* for photon energies below 1·02 MeV (the minimum energy at which pair production can occur) and above 1·02 MeV it increases with increasing photon energy (Fig. 14.8, 14.3.4).

If the photon energy is kept constant and the medium changed, κ/ρ is proportional to Z^2/A. As Z/A is approximately constant for all elements, for a given photon energy, κ/ρ is approximately proportional to Z.

14.3.6 Photo-nuclear disintegration can occur if an incident photon has an energy sufficiently high for it to cause the nucleus of an atom of the medium to disintegrate, i.e. to eject a particle such as a proton or a neutron. For the elements occurring in tissue, photo-nuclear disintegration is most likely to occur at energies in the range 20 to 25 MeV, which is above the maximum energy available in most radiotherapy centres. Even at these energies the process is not normally significant and it will not be considered further.

14.3.7 The relative importance of the individual processes is indicated in Fig. 14.8 (14.3.4). This is a graph of: (i) the attenuation coefficients for each of the processes taken separately, and (ii) the total attenuation coefficient (which is the sum of the individual coefficients) plotted against photon energy for a homogeneous beam of X rays. Classical scattering and photo-nuclear disintegration are not represented because they usually make only small contributions to the attenuation. Note that sometimes the attenuation coefficient is plotted against *wavelength* instead of photon energy (Fig. 14.10a, 14.5); the graph then appears 'left-to-right' because wavelength is *inversely* proportional to photon energy (9.2.2).

At low photon energies (long wavelengths), the total mass attenuation coefficient is large and is due almost entirely to the photoelectric process. As the photon energy increases, the photoelectric coefficient decreases rapidly and the Compton process becomes predominant. At higher energies still, the attenuation coefficient for pair production (which is zero below 1·02 MeV) increases; at very high energies it supersedes the Compton process as the main contribution to the total attenuation coefficient. Table 14.1 gives the values of the photon energies at which the mass attenuation coefficient for the photoelectric process, τ/ρ, equals that for the Compton process, σ/ρ, and the

---

*κ is the Greek small letter kappa.

TABLE 14.1 (14.3.7) *Photon energies at which the various attenuation coefficients are equal*

	Soft tissue and water	Bone and aluminium	Lead (atomic number = 82)
Photon energy at which $\tau/\rho = \sigma/\rho$	25 keV	50 keV	500 keV
Photon energy at which $\sigma/\rho = \kappa/\rho$	25 MeV	15 MeV	5 MeV

Note that the photon energies given are for *homogeneous* radiation. The *heterogeneous* radiation emitted by an X-ray set is equivalent to homogeneous radiation of photon energy numerically equal to about one-third to one-half of the peak value of the applied voltage, under the usual conditions of filtration, e.g. a set working at about 50 kVp produces heterogeneous radiation equivalent to homogeneous radiation of about 25 keV.

coefficient for the Compton process, σ/ρ, equals that for pair production, κ/ρ, in media with different atomic numbers.

The photoelectric process predominates up to higher energies in a medium of *high* atomic number than it does in a medium of *low* atomic number. This is because the mass coefficient for the photoelectric process *depends* on Z (is approximately proportional to Z^3), whereas the mass coefficient for the Compton process is almost *independent* of Z. Similarly, pair production supersedes the Compton process at lower photon energies in a medium of *high* atomic number than it does in a medium of *low* atomic number because the mass coefficient for pair production increases with Z. In consequence, the range of photon energies through which the Compton process makes the predominant contribution to the total attenuation coefficient is much smaller in a medium of *high* atomic number than it is in a medium of *low* atomic number. This can be seen by referring to the data in Table 14.1. For soft tissue, which is composed of elements of low atomic number, the Compton process predominates from 25 keV to 25 MeV, whereas for lead it predominates through the smaller range from 500 keV to 5 MeV.

14.3.8 Real absorption coefficients. The attenuation coefficients considered above relate to the *attenuation* of the beam of X or gamma radiation, i.e. to the removal of energy from the beam (14.2.2). Only a part of this energy is *truly absorbed* by the medium; it is this part which produces the effects observed in the medium, e.g. the chemical, photographic and biological changes.

Each attenuation coefficient can be divided into two parts, one relating to the energy that is truly absorbed and the other to the energy that is not absorbed. For example, for the Compton process,

$$\frac{\sigma}{\rho} = \frac{\sigma_a}{\rho} + \frac{\sigma_s}{\rho}, \qquad \text{Eq. 14.4}$$

where σ_a/ρ represents the **real** or **true mass absorption coefficient** and relates to the energy transferred to the recoil electrons, and

σ_s/ρ represents the energy scattered.

Note that when the photoelectric process occurs in a medium of low atomic number (e.g. soft tissue), the characteristic X radiation emitted, which represents the energy not absorbed in the photoelectric process, is so soft (i.e. it has such a low photon energy resulting from the low binding energies of the electrons in the shells) that it is immediately absorbed by the medium. Consequently the *attenuation* coefficient gives a measure of the energy actually absorbed by the photoelectric process in materials such as soft tissue and water.

14.4 THE TRANSMISSION OF A HETEROGENEOUS BEAM THROUGH A MEDIUM

When a heterogeneous beam of X rays passes through a uniform medium, the attenuation does *not* follow the exponential law as it does for a homogeneous beam (14.2.3). A range of photon energies is present in a heterogeneous beam and, as the value of the attenuation coefficient for a medium is not the same for all energies (Fig. 14.8, 14.3.4), different parts of the spectrum of the heterogeneous beam are attenuated by different amounts. Consequently the spectrum of the radiation changes as it passes through the medium.

For the range of energies usually encountered in radiology, where the photoelectric and Compton processes predominate, the low energy (long wavelength) end of the spectrum is attenuated to a greater extent than is the high energy end because the attenuation coefficient is greater at low energies (Fig. 14.8). Consequently, as a heterogeneous beam passes farther into a medium, not only is the beam attenuated but a larger *proportion* of the high energy photons remain (see 14.5 on filtration). The beam therefore becomes more penetrating (i.e. harder) and has a greater half-value layer (10.5.2).

Note that the *total intensity* at a point in a medium is equal to the sum of the intensity of the primary beam at the point and the intensities of all scattered, characteristic and annihilation radiations which also pass through the point. This is a particularly important consideration in radiotherapy where scattered radiation makes a significant contribution to the dose delivered at various points in a patient (14.6.2).

14.5 FILTRATION

For most heterogeneous beams (i.e. all but those containing very high energy photons only), it is possible to select a medium of such an atomic number that the attenuation coefficient increases greatly towards the low-energy end

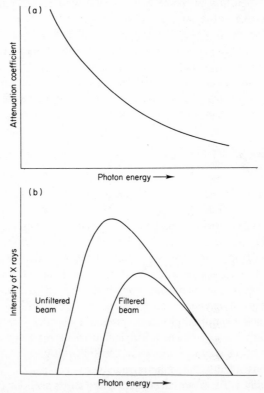

Fig. 14.10   The effect of filtration on the spectrum of a beam of X rays. (a) The attenuation coefficient of the filter medium plotted against photon energy; (b) the spectra of a beam of X rays before and after filtration.

of the spectrum of the beam (Fig. 14.10a). If the beam is passed through a **filter** comprising a layer of such a material, the low-energy (long-wavelength) photons are attenuated to ,a greater extent than the high-energy ones; consequently the low-energy radiation is filtered out of the beam. This effect is illustrated in Fig. 14.10b where the spectra of a beam before and after filtration are shown.

Filtration is said to harden a beam, i.e., a beam is more penetrating and therefore has a greater half-value layer after filtration than before. This is important in radiology because the readily absorbed low-energy radiation present in the output of an X-ray tube would result in a relatively large absorbed dose (15.4.1) to the skin of the patient if the beam were not filtered.

All X-ray tubes have some filtration, known as **inherent filtration,** due to attenuation: (a) in the target itself, (b) in the materials of the tube and of the window in the tube housing, and (c) in the cooling oil (10.4.2(iii), 11.3.4). For X rays generated below about 120 kVp, a simple filter usually made of 1 or 2 mm of aluminium is added, but from 120 kVp to about 2 MV **composite filters** (see below) made of layers of different elements must be used. The primary element chosen for the first layer of a composite filter is selected so that the attenuation coefficient increases greatly at the low photon energy end of the spectrum, as in the case of a simple filter. The elements usually chosen are:

up to 120 kVp	aluminium, simple filter	
120 kVp to 250 kVp	copper ⎫	primary element
250 kVp to 400 kVp	tin ⎬	in a composite
800 kVp to 2 MV	lead ⎭	filter.

The primary element in a composite filter must be followed by one or more layers of elements of *progressively lower atomic number.* These secondary layers are necessary because the lowest photon energies in the beam emerging from an X-ray set are less than the energy at which the K-absorption limit for the photoelectric process (14.3.3) occurs in the primary element. The latter generates characteristic X rays at energies near that of the K-absorption limit that have to be attenuated by the secondary layers. Also, a band of low-energy radiation 'leaks' through the primary layer because of the sharp drop in the attenuation coefficient at the K-absorption limit and this has to be attenuated too.

A filter with copper as the primary element is supplemented by aluminium, tin is supplemented by copper plus aluminium, and lead is

213

supplemented by tin plus copper plus aluminium. Note that the element with the *highest* atomic number must be on the side of the composite filter which is nearer to the X-ray tube. A composite filter of tin, copper and aluminium is known as a **Thoraeus filter**.

14.6   THE TRANSMISSION OF A BEAM THROUGH BODY TISSUES

14.6.1   Body tissues can be considered as being of two main types, bone and soft tissue. The essential difference between the two is that bone contains elements such as calcium ($Z = 20$) and phosphorus ($Z = 15$) that are of higher atomic number than those in soft tissue such as hydrogen ($Z = 1$) and carbon ($Z = 6$). Consequently, when the *photoelectric* process predominates, the attenuation in bone is far greater than that in soft tissue because the mass attenuation coefficient for the photoelectric process is approximately proportional to Z^3 (14.3.3). When the *Compton* process predominates, this large difference in attenuation does not occur because the mass attenuation coefficient for the Compton process is approximately independent of Z (14.3.4). If *very high* energies are involved and pair production is therefore significant, the attenuation in bone is again greater than that in soft tissues because the mass attenuation coefficient for pair production is approximately proportional to Z (14.3.5).

14.6.2   In radiotherapy the difference in the attenuation between bone and soft tissue due to the photoelectric process results in a larger absorption of energy or absorbed dose (15.4.2) in bone than in soft tissue up to (homogeneous) photon energies of about 150 keV, which corresponds to X rays generated up to about 350 kVp. If bone is traversed by the radiation when soft tissue is being treated at these energies, the bone will receive an excessively high absorbed dose. Higher photon energies are therefore employed to treat tumours underneath or near bone so that absorption in both the bone and the soft tissue is mainly by the Compton process; the mass attenuation coefficients are therefore approximately equal. Equipment which produces X rays or gamma rays of energy 1 MeV or higher is therefore used for this purpose, e.g. van de Graaff generators, linear accelerators and cobalt-60 teletherapy units. At very high energies, pair production would again result in greater absorption in bone but this is not usually significant in practice.

When calculating absorbed dose, the contribution made by scattered

214

radiation arising from the Compton process must be taken into account. The absorbed dose at a point on the surface of a patient or at a depth in the patient (the **depth dose**) has two main components, that due to absorption of the primary radiation from the incident beam and that due to absorption of the scattered radiation passing through the point. The scattered radiation is of lower energy than the primary radiation. The amount of scattered radiation increases as the Compton process becomes more predominant and it also increases with the volume of tissue irradiated, i.e. with the field-size used. The absorbed dose to be expected at each point in a patient for a known exposure is first established for the field-size and photon energy to be used. This is done by employing calibration 'phantoms' of water or other materials with attenuating properties similar to those of tissue, e.g. special waxes or 'bolus' materials. Measurements of the absorbed doses at various points in the phantom are made under the same conditions of irradiation as those to be used for the patient.

As photon energy increases, more of the radiation scattered by the Compton process is scattered in the *forward* direction, i.e. more is scattered through small angles to the direction of the incident beam and less is scattered backwards. Consequently, when X rays or gamma rays of about 1 MeV are used, instead of X rays generated at, say, 200 kVp, there is less **backscatter** to the surface of the patient and there is less radiation scattered outside the geometrical limits of the primary beam.

14.6.3 In diagnostic radiology, the image formed is basically a 'shadow picture' produced by the X rays. The attenuation of the radiation as it passes through the structure of interest depends on: (a) the mass attenuation coefficient of the structure (14.2.4), (b) the density of the structure (1.2.2), and (c) the thickness of the structure. The contrast in the image consequently depends on differences in one or more of these three factors between the structure and the surrounding tissue.

At photon energies below 25 keV (equivalent to X rays generated at about 50 kVp) the photoelectric process predominates in both bone and soft tissue (Table 14.1, 14.3.7). Consequently, bone, which contains elements of higher atomic number than soft tissue (14.6.1), has a much larger mass attenuation coefficient and the image of a bony structure produced at these energies has a high contrast.

If the photon energy is increased, the proportion of Compton interactions increases; the difference between the mass attenuation coefficients becomes less with the result that the contrast decreases. There is also an increase in the amount of scattered radiation arriving at the film or screen (see below) which further reduces the contrast. It is widely believed that Compton scatter is *wholly deleterious* to the image because of the reduction in contrast caused by some of the scattered radiation arriving with more or less uniform intensity over the area of the film or screen. However, when the predominant attenuation process *is* the Compton scatter, the contrast in the image is *produced by* the Compton interactions. It is the 'by-product' of the Compton attenuation process, viz. the scattered radiation, which may *reduce* the contrast, adversely affecting the image.

Note that the patient himself acts as a filter (14.5); the radiation emerging from the patient and reaching the film or screen has a different spectrum from that incident on the patient.

If the anatomical structure to be studied does not differ sufficiently in attenuating properties from the adjacent tissues to produce adequate contrast, it is sometimes possible to introduce into the structure a medium which is either more or less opaque to the radiation than the surrounding tissues, viz. a **contrast medium.** Commonly used contrast media which are *more* radio-opaque than soft tissue contain elements with atomic numbers considerably *higher* than those in soft tissue, e.g. preparations containing iodine ($Z = 53$) for intravenous pyelograms, and barium ($Z = 56$) for barium meal examinations. Their mass attenuation coefficients for the photoelectric process are consequently very high. Examples of contrast media which are *less* radio-opaque than soft tissue are gases, e.g. air or carbon dioxide for techniques such as ventriculography. Their *densities* are much *lower* than that of soft tissue.

If an appreciable amount of scattered radiation reaches the film or screen, it reduces the contrast of the image by increasing the general background. Radiation is scattered (by the Compton process) in the patient and also in any nearby object which is irradiated; the effect can be reduced by:

(i) using diaphragms and cones to limit the size of the beam and hence to restrict the volume of tissue irradiated,

(ii) placing an absorbing material, usually lead, behind the film-screen combination in the cassette to reduce backscatter from objects behind the cassette,

(iii) using a compression band or cone to reduce the volume of tissue in the field,

(iv) using a grid (see below) which transmits radiation travelling in the direction of the primary beam but absorbs scattered radiation travelling in other directions,

(v) increasing the distance between the patient and the film **(exit-gap or air-gap technique)** so that less radiation scattered by the patient reaches the film.

Grids are made of narrow strips of lead, about 0·1 mm thick, 5 mm high and spaced about 0·5 mm apart. The strips are arranged parallel to one another and placed between the patient and the film. The strips are inclined at an angle to the plane of the film so that X rays radiating from the focus of the tube can pass between the strips; scattered radiation travelling in other directions, however, is intercepted by the strips and prevented from reaching

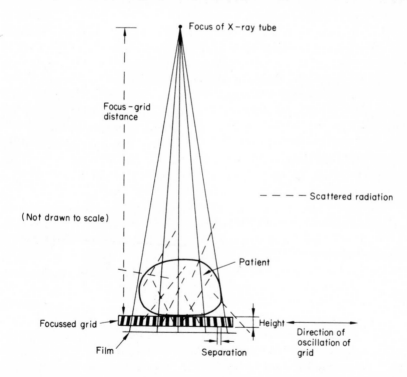

Fig. 14.11 The use of a focussed grid to reduce the amount of scattered radiation reaching the film.

the film (Fig. 14.11). A grid with inclined strips is referred to as a **focussed grid**; it is designed to be used at a particular focus-grid distance. The effectiveness of a grid is indicated by the **grid ratio** which is the ratio of height to separation of the lead strips. Grid ratios are usually in the range 5:1 to 16:1, the higher ratio grids being used for higher energy radiation.

A grid as described above would produce a shadow on the film. In order to prevent this image of the grid from interfering with the image of the structure being radiographed, the grid is made to oscillate during the exposure in a plane parallel to the film and in a direction at right-angles to the strips, as in a **Potter-Bucky diaphragm**.

14.7 SHAPES AND FINE DETAILS IN THE X-RAY IMAGE

14.7.1 The **shapes** of the patterns in the 'shadow picture' (14.6.3) formed by the X rays that emerge from the patient (sometimes called the **X-ray** or **radiation image**) depend on the shapes of the structures that produce the shadows. It is from the shapes of these shadows, detected by an image intensifier or by a screen–film combination, that the radiologist can acquire information about the internal structures of the patient. However, because the structures are three-dimensional, and the patterns only two-dimensional, the shapes of the patterns may be distorted in a way that is reminiscent of the distortion in land shapes caused by representing the spherical earth on flat maps. The radiologist must be aware of this distortion and must allow for it in his diagnosis.

The X-ray beam *diverges* from the focal area of the X-ray tube, hence the images of body structures must be larger than the structures themselves. This is called **radiological magnification** or **enlargement** (Fig. 14.12a). Using the principle of similar triangles, Fig. 14.12a shows that the ratio of image to object sizes is equal to the ratio of focus-image to focus-object distances:

$$\frac{x_i}{x_o} = \frac{FI}{FO}.$$

Eq. 14.5

Radiological magnification may be a disadvantage, for example in chest radiography when an exact estimate of heart size is required. Then FI is made as nearly as possible equal to FO by positioning the patient in contact with the film cassette and by having the X-ray tube at an unusually large distance from both. On the other hand, radiological magnification may be deliberately

218

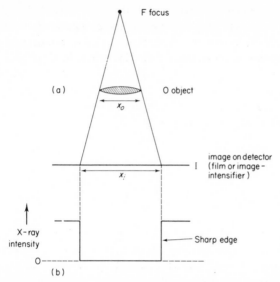

Fig. 14.12 (a) Illustrating radiological magnification or enlargement. FO and FI are the focus-object and focus-image distances respectively. (b) Variation of X-ray intensity with distance. This shows how a point source of X rays produces an image with sharp edges. The object is assumed to have clear-cut (well-defined) edges and to absorb completely the X rays incident on it, so that the X-ray intensity beneath it is zero.

made use of as in the technique called **macroradiography**, used for biological specimens as well as for patient structures, when FO is made much smaller than FI. In this case, however, a very small effective focal spot size (11.3.3) is necessary to reduce the effects of geometrical unsharpness (14.7.2).

14.7.2 The **fine details** in the X-ray image result not only from the fine structure in the object but depend also on the effective focal-spot size (11.3.3) of the X-ray tube. This can be demonstrated by a simple optical experiment. If an object with a clear-cut edge, such as a piece of thin card, casts a shadow from a very small (point) light source such as an electric torch bulb (minus its reflector!), the edge of the shadow will also be clear-cut or **sharp**. Fig. 14.12b shows similarly how an ideal point of source of X rays would produce a sharp image from an object with clear-cut edges. If, instead, a light source of large area, such as an ordinary pearl electric light-bulb, is used, the edge of the shadow will be **blurred**, or **unsharp**. The phenomenon is called **blurring**, or **unsharpness**.

Exactly the same effect occurs with the finite area of the effective focal-spot of an X-ray tube; the shadows of 'sharp' objects are rendered unsharp (11.3.3). In the X-ray case the phenomenon is called **geometrical unsharpness**. Fig. 14.13 shows in more detail how this arises. In the region of complete

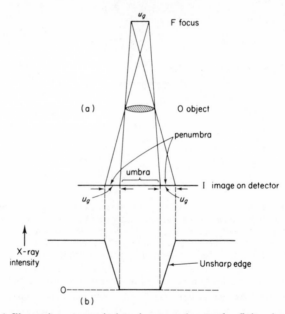

Fig. 14.13 (a) Illustrating geometrical unsharpness due to the finite size of the X-ray source. u_g is the effective focal-spot size and U_g is the size of the resultant image unsharpness or penumbra. (b) Variation of X-ray intensity with distance. The object is assumed to have clear-cut edges and to absorb the X rays completely.

shadow (the **umbra**), no X rays fall on the detector. In the region of partial shadow (the **penumbra**), X rays from only part of the focal area fall on the detector. The fraction of X-ray energy that reaches the detector at a given point in the penumbra depends on the distance of the point from the edge of the umbra. This results in a variation of intensity at the edge of the image that gives it a blurred or unsharp appearance (Fig. 14.13b).

The extent of geometrical unsharpness, as with radiological magnification, depends on the relative positions of focus, object and detector (film, etc.). Fig. 14.13a shows that

$$\frac{U_g}{u_g} = \frac{IO}{FO}$$
Eq. 14.6

(by similar triangles). The effect (U_g) of a given effective focal-spot size (u_g) is therefore least when the object is as near the detector as possible. This is why the technique of macroradiography (14.7.1) requires a very small effective focal-spot size. Further information about focal-spot sizes will be found in section 11.3.3.

There are two other sources of unsharpness in the X-ray image. First, if any of the three components, X-ray source, object or film, moves relative to the others during the exposure, an unsharp or blurred image will result. This is similar to the unsharpness experienced when a rapidly-moving object such as a racing car is photographed using a camera set for too long an exposure time. The effect is called **movement unsharpness** (U_m), and can be minimized only by reducing the exposure time. This reveals the advantage afforded by big X-ray generators capable of producing large values of current (mA), because then the same current x time product (mAs) can be obtained with a shorter exposure time. Normally, of course, it is desirable to minimize movement unsharpness, but in a valuable technique called **tomography**, by suitable relative movement of X-ray tube and film, gross movement unsharpness can be introduced into all planes of the patient except the one of interest, thus rendering the structures in that plane more or less clearly visible.

Apart from tomography, the types of unsharpness so far discussed have been undesirable. However, another type, called **absorption unsharpness** or **structure unsharpness** (U_a) results from the fact that some body structures, for example eroded bone, do not have clear-cut edges. The resulting unsharpness in the X-ray image is then a desirable feature, because it gives information about the fine structure of the edges of objects. The radiographer must not assume that an unsharp edge in a radiograph is necessarily a bad thing, to be eliminated at all costs!

So far we have discussed effects that give rise to unsharpness in the X-ray image itself. However, the X-ray image is not visible, and has to be rendered so by means of a screen—film combination or an image-intensifier television system. These devices, which may be called X-ray image **transducers**, always introduce extra unsharpness in converting X rays into light. Most X-ray transducers contain a fluorescent screen (9.3.5) as one of their essential components, and so this type of unsharpness is traditionally known as **screen unsharpness** (U_s). In the more complex types of transducer used today,

221

however, unsharpness arises also in other parts of the system, for example in the television part of an image-intensifier system, and the theory is extremely complex. Furthermore, the simple description of unsharpness in terms of a distance in mm has for many purposes become inadequate, often being replaced by a mathematical concept called the **modulation transfer function**.

It is obviously advantageous for 'screen' unsharpness to be a minimum; this is achieved in radiography by using **high-definition intensifying screens** (9.3.5) (at the expense, however, of sensitivity), and in fluoroscopy by choosing the most modern image-intensifying system, for example, one incorporating the so-called **caesium iodide** intensifier tube (9.3.5). Another way of reducing the undesirable effects of screen unsharpness is to use the technique of macroradiography (see above). In this way the image of the object of interest, together with its fine details, is made larger, thus rendering the fixed size of the screen unsharpness less important relative to the size of the X-ray image details.

In a radiological system there are many sources of unsharpness, as described above. It is often desired to calculate the **total unsharpness** (U_t) resulting from the combination of the several individual unsharpnesses. The only exact way to do this is by a mathematical technique making use of the **modulation transfer function** (see above), but an approximate result may be obtained by the following simple equation. For example, the result of combining geometrical unsharpness (U_g), movement unsharpness (U_m) and screen unsharpness (U_s) is given by

$$U_t = \sqrt{U_g{}^2 + U_m{}^2 + U_s{}^2}. \qquad \text{Eq. 14.7}$$

15 X-Ray and gamma-ray measurements

15.1 THE BASIS OF THE MEASUREMENTS

The measurement of quantity of X or gamma radiation can be based on any of the effects produced by the radiation, e.g. ionization, chemical changes and photographic effects. Some effects are readily measurable only if large quantities of radiation are involved and some depend very much on the quality of the radiation producing them. Consequently the use which can be made of these effects is limited.

The effect which is most suitable for general use over a wide range of qualities is the ionization (2.4) produced in air by the radiation. This measure of quantity is known as **exposure** (15.2.1). The amount of ionization produced is related to the energy absorbed by the air (14.3.8) and is measured by collecting the ions formed, as described in section 15.2.

Much of the early interest in the measurement of X-ray quantity was in connexion with the doses given to soft tissue in radiotherapy. X-ray quantity expressed in terms of the ionization of air is particularly useful for this purpose as it gives a measure of the energy absorbed by soft tissue exposed to radiation. This is because air and soft tissue have approximately equal effective* atomic numbers (7.6 and 7.4 respectively) and hence similar attenuation coefficients (14.3); therefore the energy absorbed per unit mass is the same for *equal exposures.*

---

* The effective atomic number of a mixture of elements is a kind of average of the atomic numbers of the individual elements.

15.2 EXPOSURE

15.2.1 Exposure and the roentgen. Quantity of X or gamma radiation measured in terms of the ionization produced in air is known as **exposure**.

DEFINITION   **Exposure is the quantity of X or gamma radiation measured as the sum of the electric charges on all the ions of one sign produced in air when all the secondary electrons liberated by the photons in a small mass of air are completely stopped in air.**

The detailed conditions given in the definition will be explained in section 15.2.2.

The unit of exposure is the **roentgen (R)**.

DEFINITION   **The roentgen is an exposure of $2\cdot58 \times 10^{-4}$ coulomb/ kilogram of air.***

Note that exposure and the roentgen relate only to X and gamma radiation which are both electromagnetic. Alpha and beta radiations are measured in terms of absorbed dose in a stated medium (15.4.1).

The roentgen is not an S.I. unit and its use may be discontinued in the future.

15.2.2 The free-air ionization chamber is an instrument for making an accurate and absolute measurement of exposure; it satisfies the conditions in the definitions for exposure and for the roentgen given in section 15.2.1. It is a cumbersome instrument not found in hospital departments and is used for calibration purposes in national standardizing laboratories.

A section through the free-air chamber is shown in Fig. 15.1. The X rays enter the lead-lined box through a small hole in the lead diaphragm which produces a well-defined conical beam. The beam passes between two parallel plate electrodes which have a high potential difference across them and which collect (by attraction) some of the ions produced in the air in the box. The bottom electrode is divided into two parts, the collecting plate C and the 'guard' ring G around C. The guard ring G and the series of guard wires W help to keep the lines of electric force between the collecting plate C and the

---

* Prior to 1962, the roentgen was defined as the quantity of X or gamma radiation such that the associated corpuscular emission (i.e. the secondary electrons liberated by the photons) per $0\cdot001\ 293$ gram of air produces, in air, ions carrying one electrostatic unit (3.3.1) of quantity of electricity (i.e. electric charge) of either sign. The two definitions are equivalent.

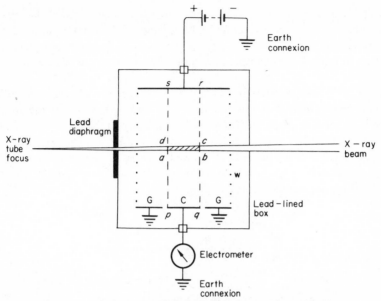

Fig. 15.1   A cross-section through the free-air ionization chamber.

opposite electrode straight and perpendicular to the electrodes; this is done in order to define accurately the volumes involved in the measurement.

Consider the secondary electrons liberated by X-ray photons interacting with the small mass of air in the volume represented by *abcd*; this is bounded by the edges of the beam and the lines of electric force *ps* and *qr* from the edges of the collecting plate C. The plate electrodes are sufficiently far apart for the secondary electrons not to reach them and so the electrons give up their energy by producing ions *in air*. The potential difference between the electrodes is sufficiently high for all the ions of one sign produced in the volume represented by *pqrs* to be collected on plate C (i.e. no recombination of ions occurs). Their charge is then measured on the electrometer (3.2.1) connected to plate C. Some of the electrons liberated by the X rays in the volume *abcd* will, however, pass out of the collecting volume *pqrs* and produce ions outside it; these will not be collected on plate C. This loss of ions, however, is compensated for by ions produced in the collecting volume *pqrs* by secondary electrons liberated by X-ray photons interacting with the air *outside* volume *abcd*. This compensation is known as **electronic equilibrium**.

The exposure in roentgens can then be calculated from the *charge* measured on the electrometer and a knowledge of the mass of air in volume *abcd*. If the *current* (i.e. the charge per second) flowing from plate C is measured instead, the **exposure rate** in roentgens/second can be calculated.

15.2.3 Thimble ionization chambers are used for clinical and other applications where it is not practicable or appropriate to use the free-air chamber because of its great size and complexity (15.2.2). In the thimble chamber, a small volume of air is surrounded by a solid wall which is **air-equivalent**, i.e. the material of the wall is chosen so that it has an effective atomic number approximately equal to that of air, and consequently a similar mass absorption coefficient (Fig. 15.2).

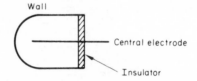

Fig. 15.2   A thimble ionization chamber.

The wall is sufficiently thick for secondary electrons liberated *outside* the chamber not to penetrate it. Consequently when the chamber is exposed to X or gamma radiation, the ionization produced in the air inside the chamber is due mainly to secondary electrons liberated in the wall and passing through the air in the chamber. There is also a small contribution due to ionization resulting from absorption of the X or gamma radiation by the air in the chamber. The ions are collected by applying a potential difference between the central electrode and the wall of the chamber, the latter having an electrically conducting coating on its inner surface. The potential difference must be sufficiently high to ensure that all the ions produced are collected, i.e. that there is no recombination of ions, and that therefore the ionization chamber is operating in a saturated condition (11.1.3).

If the mass of air enclosed in the chamber is increased (e.g. by varying the volume of the chamber, or the pressure or the temperature of the air), there is a greater mass of air in which the secondary electrons can produce ions. Consequently the charge collected is greater, i.e. the sensitivity of the chamber is increased.

Thimble ionization chambers do not fulfil the conditions in the definitions of exposure and of the roentgen (15.2.1), hence the ionization is not *equal* to the exposure. However, as the walls are air-equivalent, these chambers can be made so that the ionization produced is *proportional* to the exposure in roentgens over a range of radiation qualities. The constant of proportionality for a given thimble chamber is obtained by calibrating it against a free-air chamber for the quality of radiation to be measured. Care must be taken to ensure that the wall is of the proper thickness for the energy of the radiation being measured. If the wall is too thin, insufficient absorption takes place in the wall and too few secondary electrons are liberated. Also, electrons from outside the chamber can pass through the wall. If the wall is too thick, there will be appreciable attenuation of the X or gamma radiation in the outer layers of the wall resulting in reduced ionization in the air in the chamber. If a chamber with a thin wall is to be used for measurements at energies higher than that for which it was designed, the wall can be increased to the required thickness by surrounding the chamber with a **build-up cap** such as a Perspex sheath a few millimetres thick.

15.2.4 Exposure meters and exposure-rate meters. A thimble ionization chamber is connected by an electrically screened cable to an instrument (an electrometer) which measures the charge collected on the central electrode; the total exposure in roentgens is obtained if the chamber has previously been calibrated (15.2.3). In this arrangement the chamber is used as an **exposure meter**. Alternatively, the chamber can be connected to an instrument which measures the ionization current (i.e. the charge per second) flowing through the central electrode; thus the exposure rate in roentgen/second is obtained, i.e. the chamber is used as an **exposure-rate meter**.

Sometimes it is inconvenient to measure exposure using a chamber which is connected to an electrometer by a cable during the irradiation; instead, a **condenser ionization chamber** (Fig. 15.3) is used which is not connected to the electrometer during the irradiation. In use, the screwed plug in the

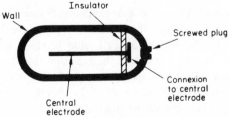

Fig. 15.3   A condenser ionization chamber.

chamber wall is removed and the central electrode is charged from a charging device. The plug is replaced and the chamber exposed. The plug is removed again and the charge remaining on the central electrode is measured by connecting it to an electrometer.

The chamber itself acts as a capacitor (or condenser, 3.4.2). The chamber is charged initially to a chosen potential difference V_1 between the central electrode and the wall. This is related to the initial charge Q_1 and the electrical capacitance of the chamber C by substituting in Eq. 3.1 (3.4.1).

Hence

$$Q_1 = CV_1.$$

During the exposure of the chamber, the ions collected on the central electrode neutralize some of the charge. If V_2 represents the potential difference measured between the central electrode and the wall after exposure, the charge remaining Q_2 is given by:

$$Q_2 = CV_2.$$

Hence, the charge collected on the central electrode during the exposure $(Q_1 - Q_2)$ is given by:

$$(Q_1 - Q_2) = C(V_1 - V_2).$$

Thus the measured fall in potential difference $(V_1 - V_2)$ is proportional to the charge collected and, with the calibration factor appropriate to the chamber, the value of the exposure in roentgens can be calculated.

An example of a condenser ionization chamber is the Medical Research Council chamber type BD11; this has a number of applications, including the measurement of exposure in radiological protection work (17.5.3).

15.3 THE MEASUREMENT OF HALF-VALUE LAYER

The half-value layer of a beam of X or gamma radiation is that thickness of a stated material which reduces the exposure rate of the beam to one-half (10.5.2, 14.2.5). To determine the half-value layer of a beam, a thimble ionization chamber is used to measure the exposure rate. The radiation passes through an aperture in a lead diaphragm to produce a narrow beam which is just broad enough to irradiate the whole of the chamber. The exposure rate is measured first with no attenuating material present and then with various thicknesses of the material placed over the aperture on the side of the diaphragm nearer the X-ray tube. A transmission curve of exposure rate against thickness of the material is plotted and the half-value layer is read off the graph as the thickness required to reduce the exposure rate to one-half.

15.4 ABSORBED DOSE

15.4.1 Absorbed dose and the rad. When a medium is traversed by radiation, energy is imparted to the medium and the effects produced in the medium result from this *absorption* of energy (14.1.1). The amount of this absorption can be expressed in terms of the **absorbed dose.**

DEFINITION   **Absorbed dose is the energy imparted by the radiation to the medium per unit mass of the medium.**

The unit used for absorbed dose is the **rad.**

DEFINITION   **The rad is an absorbed dose of $\frac{1}{100}$ joule/kilogram of the stated medium.**

Unlike exposure and the roentgen, which apply to X and gamma radiation only, absorbed dose and the rad apply to alpha and beta radiation as well as to X and gamma radiation.

The rad, being an absorbed dose of only 1/100 joule/kilogram of medium, is not itself an S.I. unit. It is being replaced by a new S.I. unit called the **gray (Gy)** which is defined as an absorbed dose of 1 joule/kilogram of the stated medium. Note that 1 gray = 100 rad.

15.4.2 The relationship between the rad and the roentgen. The energy absorbed by a medium traversed by X or gamma radiation depends on the value of the real mass absorption coefficient for the medium at the energy of the radiation (14.3.8). Consequently, when materials with different atomic numbers are exposed to the same amount of radiation, e.g. to 1 roentgen, the amounts of energy absorbed will in general be different, because the value of the mass attenuation coefficient varies with atomic number (14.3.7). This is illustrated in Table 15.1 where the absorbed doses in air, water, soft tissue and bone exposed to 1 roentgen of monochromatic X or gamma radiation are given for three different photon energies. At 10 keV, for example, the absorbed dose in bone is approximately four times that in soft tissue. This is because the absorption in bone and in soft tissue at this energy is largely due to photoelectric absorption for which the mass attenuation coefficient is

TABLE 15.1 (15.4.2)   *Absorbed doses in various materials exposed to 1 roentgen of monochromatic X or gamma radiation*

Photon energy	Absorbed dose in rad			
	Air	Water	Soft tissue	Bone
10 keV	0·87	0·91	0·93	3·6
100 keV	0·87	0·95	0·95	1·5
1 MeV	0·87	0·97	0·96	0·92

approximately proportional to the atomic number cubed (14.3.3). Consequently, at this energy, bone has a much larger mass absorption coefficient and hence receives a much larger absorbed dose than soft tissue because bone has a higher effective atomic number than soft tissue.

At 1 MeV, however, the Compton process predominates and, because the mass attenuation coefficient for this process is approximately independent of atomic number (14.3.4), there is little difference between the mass absorption coefficients for bone and soft tissue at this energy. Consequently the absorbed doses are approximately equal.

This difference in the absorbed dose for the same exposure is important in radiotherapy. If low energy radiation were used to treat soft tissue adjacent to or containing bone, the bone would receive a much higher absorbed dose than the soft tissue, with consequent danger of bone damage. Because of this, high energies are often used in treatment so that the Compton process predominates and the absorbed doses in bone and soft tissue are approximately equal for the same exposure.

15.5 OTHER METHODS OF MEASUREMENT

15.5.1 Indirect measurements of exposure or exposure rate can be made using instruments which do not satisfy the conditions set out in the definitions of exposure and the roentgen (15.2.1); these therefore must be calibrated for the quality of radiation to be measured in order to obtain a value for the exposure. Six commonly used types of instrument are described in the following sections.

15.5.2 Photographic film dosemeters make use of the photographic effect of radiation. Their commonest form is the film badge used for personnel dosimetry in radiation protection work; it is described in section 17.5.2.

15.5.3 Thermoluminescent dosemeters make use of the property of thermoluminescence possessed by certain crystalline powders such as lithium fluoride. In thermoluminescence, the crystals store the energy absorbed during their exposure to radiation and subsequently release it as light when heated. The amount of light emitted is measured and can be related to the exposure in roentgens if the crystals and the measuring instrument are calibrated. Thermoluminescent dosemeters can be made quite small and are used in a number of applications including radiation protection work (17.5.4).

15.5.4 Geiger–Müller counters, sometimes called Geiger counters or G.M. counters, are sensitive radiation detectors which comprise a **Geiger tube** and associated electronic circuits for displaying the output of the tube on a meter or other device. The Geiger tube resembles a thimble ionization chamber but is filled with an inert gas (argon) at reduced pressure instead of air at atmospheric pressure. A high p.d. (from 400 to 1500 V) is maintained across the electrodes. When ionizing radiation produces a secondary electron in the tube, the electrons produced by ionization of the gas acquire such high energy from the high p.d. that they in turn ionize other gas molecules; this is known as **gas amplification.** The result is cumulative ionization (a chain reaction) which produces a large negative pulse containing a very large number of electrons (11.2.3). A series of such pulses can be counted automatically by electronic circuits.

Geiger counters are more sensitive than ionization chambers because of the gas amplification; they are widely used as monitoring instruments for low intensities of radiation. They do not satisfy the conditions in the definitions of exposure and the roentgen and therefore have to be calibrated before they can be used to measure exposure or exposure rate.

15.5.5 Scintillation counters make use of the fact that when certain solids and liquids, called **phosphors,** are exposed to radiation, they absorb energy and then scintillate or fluoresce (9.3.5), i.e. they re-emit the energy as photons of visible or ultra-violet light. The phosphor is mounted so that the light emitted enters the window of a photomultiplier tube (an especially sensitive form of photoelectric cell, 13.4.2) which converts the photons of light into pulses of electric charge; these are then amplified and counted in the associated electronic circuits.

15.5.6 Solid-state detectors can be divided into two groups: (i) the thermoluminescent (and similar) materials described in section 15.5.3 and (ii) certain semi-conductor materials which have properties similar to those utilized in the transistor and the solid-state diode (11.3.2). The latter type of material can be used in a small robust probe which, when connected to suitable electronic circuits, can be calibrated to measure exposure rate.

15.5.7 Chemical dosemeters make use of the chemical changes produced by radiation in the material through which it passes (14.1.1). The most widely used is known as the **Fricke dosemeter** in which a solution of ferrous sulphate is exposed. The ferrous sulphate is oxidized to ferric sulphate and the amount of ferric sulphate produced, which is determined by chemical techniques, gives a measure of the exposure.

16 Radioactivity

16.1 THE DISCOVERY OF RADIOACTIVITY

16.1.1 Radioactivity. We noted previously (2.3) that an atomic nucleus having a certain proportion of neutrons to protons is *unstable*, i.e. it is **radioactive**. In such a nucleus, a transformation or disintegration occurs *spontaneously*, usually with the emission of a particle and sometimes also with the emission of a gamma-ray photon. In the process, an atom of another nuclide is formed; this is known as the **daughter product** of the original or **parent** nuclide. This process of transformation is called **radioactivity**.

DEFINITION   **Radioactivity is the property possessed by certain nuclides of undergoing spontaneous transformation of their nuclei accompanied by the emission of particles or radiation.**

As radioactivity is a property of the *nucleus*, it is unaffected if radioactive atoms enter into chemical combination or if the atoms are subjected to physical changes such as an increase in temperature.

Nuclides (isotopes) which are radioactive are known as **radionuclides (radioisotopes)**.

16.1.2 The discovery of natural radioactivity. Naturally occurring radioactivity was discovered independently in 1896 by A.H. Becquerel and by S.P. Thompson. Becquerel knew that fluorescence (9.3.5, 10.1(i)) was associated with the X rays discovered by Röntgen in 1895 and this led him to investigate the various uranium compounds that fluoresce in sunlight. He discovered that these compounds emitted penetrating rays even in the dark when they were not fluorescing. These rays could blacken photographic plates in light-tight wrappings, discharge electroscopes, and cause zinc sulphide to fluoresce.

Other workers showed that thorium compounds were similarly 'radioactive' Later, Mme Curie, realizing that crude pitchblende ore was more active than could be accounted for by its uranium content, discovered two further naturally occurring radionuclides: polonium (which she named after her native Poland) and radium.

16.1.3   The uranium-radium series. The radium isotope, $^{226}_{88}Ra$, is the naturally occurring radionuclide most widely used in medicine. It is a member of one of three chains or **series** of radionuclides occurring in nature, in which an atom of one nuclide disintegrates to form an atom of another nuclide, which is itself radioactive. This atom in turn disintegrates to form an atom of yet another radionuclide and so on until a *stable* nuclide is reached. (Several other radionuclides occur naturally which are *not* members of the three series, e.g. $^{40}_{19}K$, a radioactive isotope of potassium, 16.4.)

The first member of the series which contains radium is an isotope of uranium; the series proceeds from uranium through fourteen radionuclides until a stable isotope of lead is formed. Radium is the sixth member of the series. It has a half-life (16.5.2) of 1620 years and disintegrates with the emission of an alpha particle (16.3.1) to form the radioactive gas radon. This in turn disintegrates to a radioactive isotope of polonium and so on through the series. The radium used in medicine is *sealed* in a container, e.g. a radium needle; consequently the other members of the series after radium are present in the container too. The members of the series are said to be **in equilibrium** with the radium and they all contribute to the total radiation emitted by the sample. The wall of the container is of such a thickness, however, that only the gamma radiation (16.3.3) penetrates it and emerges from the radium needle for use in radiotherapy.

16.2   TYPES OF RADIATION EMITTED

It was discovered that three types of radiation were emitted by the various naturally occurring radioactive substances:

(i) **Alpha particles** (α particles) Some of the rays emitted were deflected by electric and magnetic fields (3.1.2, 6.2.2) in a manner that showed them to be streams of particles, each particle having a mass of 4 atomic mass units and an electric charge of +2 atomic units. These particles, called alpha particles, were easily stopped, even by a thickness of a few sheets of paper or tens of millimetres of air. It was eventually shown that an alpha particle consists of 2 protons plus 2 neutrons and is therefore the nucleus of an atom of helium (2.1.3).

(ii) Beta particles (β particles). Some of the other rays emitted were deflected by electric and magnetic fields in a manner which showed that they also were streams of particles, but in this case each particle had a mass of only $\frac{1}{1840}$ atomic mass unit and an electric charge of -1 atomic unit. These particles, called beta particles, were more penetrating than alpha particles and a few millimetres of aluminium were necessary to stop them. It was eventually shown that a beta particle is a fast-moving electron.

(iii) Gamma rays (γ rays). Other rays were *not* deflected by electric and magnetic fields and therefore were *not* streams of electrically charged particles. These rays, called gamma rays, were very penetrating and it took several tens of millimetres of lead to reduce their intensity by an appreciable amount. It was eventually shown that gamma rays are electromagnetic radiation of short wavelength (9.1.3). They are identical in kind to X rays except for their manner of production (16.3.3).

16.3 TRANSFORMATION PROCESSES

16.3.1 Alpha-particle emission. In the case of radionuclides which emit alpha particles, a disintegrating nucleus ejects 2 protons plus 2 neutrons as one alpha particle. The daughter product (16.1.1) so formed has an atomic number which is 2 less than that of the parent nuclide and a mass number which is 4 less.

This **radioactive decay process** is illustrated in Fig. 16.1, which represents the **decay scheme** of the isotope of radon, $^{222}_{86}$Rn. The parent nuclide has an atomic number of 86 and a mass number of 222. An alpha particle of energy

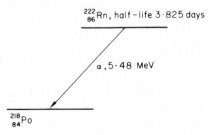

$^{222}_{86}$Rn, half-life 3·825 days

α, 5·48 MeV

$^{218}_{84}$Po

Fig. 16.1 An example of a radionuclide which emits an α particle. The atomic number decreases by 2 and the mass number decreases by 4.

235

5·48 MeV is emitted by the nucleus to form the daughter product $^{218}_{84}$Po which is an isotope of polonium having an atomic number of 84 and a mass number of 218. By convention, when the atomic number decreases, the transformation is represented as in Fig. 16.1 by a line drawn downwards and to the left from the parent to the daughter product.

16.3.2 Beta-particle emission. Beta particles are now known to be electrons with either a negative or a positive electric charge. The electron normally encountered, e.g. in the shells of an atom (2.1,3), is known as a **negatron** and has a negative charge of 1 atomic unit and a mass of $\frac{1}{1840}$ atomic mass unit. A positively charged electron, known as a **positron**, has the same mass as a negatron but has a *positive* charge of 1 atomic unit. It is only encountered in the special circumstances of radioactivity and of pair production (14.3.5).

For a beta particle to be ejected from a nucleus undergoing a radioactive transformation, either a neutron or a proton in the nucleus must undergo a change which results in the formation of the beta particle. For radionuclides which eject a negatron, a neutron changes into a proton plus a negatron, i.e.

$$_0n \rightarrow {_+p} + {_-e}$$

where $\quad$ $_0n$ is a neutron with charge 0,

$\quad\quad\quad\quad\quad$ $_+p$ is a proton with charge +1,

$\quad\quad\quad\quad\quad$ $_-e$ is a negatron with charge −1.

The positive charge on the nucleus (i.e. the number of protons) increases by 1 atomic unit because there is one more proton present in the nucleus after the transformation than before. Consequently the atomic number of the daughter product is 1 greater than that of the parent. The mass of the nucleus remains approximately the same, because the negatron has little mass, and the mass number is unchanged because the total number of protons plus neutrons is unaltered.

For radionuclides which eject a positron, a proton in the nucleus changes into a neutron plus the positron, i.e.

$$_+p \rightarrow {_0n} + {_+e}.$$

The positive charge on the nucleus (i.e. the number of protons) decreases by 1 atomic unit and so the atomic number of the daughter product is 1 less than that of the parent. The mass of the nucleus remains approximately the same and the mass number is unchanged because the total number of protons plus neutrons is unaltered.

An example of a radionuclide which emits a negatron (β^- particle) is given in Fig. 16.2 which represents the decay of $^{45}_{20}$Ca, an isotope of calcium, to $^{45}_{21}$Sc, an isotope of scandium. An example of the emission of a positron (β^+ particle) is given in Fig. 16.3 which represents the decay of $^{15}_{8}$O, an isotope of oxygen, to $^{15}_{7}$N, an isotope of nitrogen.

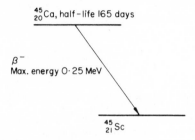

Fig. 16.2   An example of a radionuclide which emits a β^- particle. The atomic number increases by 1 and the mass number remains unchanged.

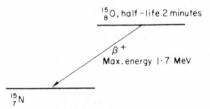

Fig. 16.3   An example of a radionuclide which emits a β^+ particle. The atomic number decreases by 1 and the mass number remains unchanged.

Note that for a given radionuclide, the beta particles emitted from the individual nuclei do not all have the same amount of kinetic energy. Their energies are spread over a range of values up to a definite maximum for the particular radionuclide, i.e. the beta particles have a continuous spectrum of energy. It is the maximum energy in the spectrum that is given in the decay schemes in Figs 16.2 and 16.3.

16.3.3   Gamma-ray emission. In the case of some radionuclides, one or more gamma-ray photons are emitted by the nucleus following the ejection of a particle because the nucleus still possesses an excess of energy.

An example of a beta-gamma emitter is given in Fig. 16.4. The isotope of gold, $^{198}_{79}$Au, decays with the emission of a β^- particle to an isotope of

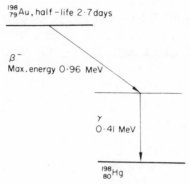

Fig. 16.4   An example of a radionuclide which emits a β^- particle followed by a γ-ray photon. (Simplified version of the decay scheme for $^{198}_{79}$Au.) The emission of the β^- particle results in an increase of 1 in the atomic number and no change in the mass number. The emission of the γ-ray photon affects neither the atomic number nor the mass number.

mercury, $^{198}_{80}$Hg. The nucleus of the $^{198}_{80}$Hg formed by the ejection of the β^- particle has an excess of energy; this energy is then emitted as one gamma-ray photon of energy 0·41 MeV.

Note that when a gamma-ray photon (electromagnetic radiation) is emitted, there is no further change in atomic number or mass number. Note also that the gamma-ray photons from a given radionuclide are emitted with a definite value or values of energy characteristic of that radionuclide and not with a continuous spectrum of energy as is the case for beta particles. Some radionuclides, for example $^{198}_{79}$Au, emit only one gamma-ray photon in their decay schemes. Others emit more than one gamma-ray photon; for example, $^{60}_{27}$Co, the radioisotope of cobalt used in teletherapy units, emits two gamma-ray photons in its decay scheme, one of energy 1·17 MeV and the other of energy 1·33 MeV.

16.3.4   The electron-capture process. The electron-capture or K-capture process also results in a transformation of the nucleus although no particle is ejected from it. Instead, the nucleus captures an electron from the K-shell of the atom and the electron combines with a proton in the nucleus to form a neutron, i.e.

$$_-e + {}_+p \rightarrow {}_0n$$

where $_-e$ is the electron,

                          $_+p$ is the proton,

                          $_0n$ is the neutron.

The positive charge on the nucleus (i.e. the number of protons) decreases by 1 atomic unit and so the atomic number decreases by 1. The mass of the nucleus remains approximately the same and the mass number is unchanged because the total number of protons plus neutrons is unaltered. In some cases, the vacancy produced in the K-shell is filled by the transition of an electron from another shell with the emission of a characteristic X-ray photon (10.3(iii)).

The electron capture process is illustrated in Fig. 16.5 which represents the transformation of an isotope of argon, $^{37}_{18}$A, to an isotope of chlorine, $^{37}_{17}$Cl.

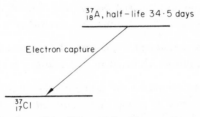

Fig. 16.5 An example of a radionuclide which undergoes electron capture. The atomic number decreases by 1 and the mass number remains unchanged.

16.4 BRANCHING

Certain radionuclides can decay by more than one process, that is, fixed proportions of the number of nuclei disintegrating in a sample decay by

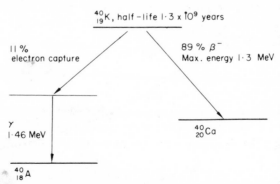

Fig. 16.6 An example of a radionuclide which decays by more than one process. Eleven per cent of the transformations are by electron capture followed by the emission of a γ-ray photon to $^{40}_{18}$A. Eighty-nine per cent of the transformations are by the emission of a $β^-$ particle to $^{40}_{20}$Ca.

239

different processes. Fig. 16.6 illustrates the decay scheme for the naturally occurring radioactive isotope of potassium, $^{40}_{19}K$. The scheme has two **branches:** 11% of the transformations are by electron capture followed by the emission of a gamma-ray photon to form an isotope of argon, $^{40}_{18}A$, while the remaining 89% of the transformations are by the emission of a β^- particle to form an isotope of calcium, $^{40}_{20}Ca$.

16.5   RADIOACTIVE DECAY

16.5.1   Activity and exponential decay. The rate at which a sample of a radionuclide is undergoing transformation is known as its **activity.**

DEFINITION   **The activity of a quantity of a radionuclide is the number of nuclear transformations which occur in that quantity per unit time.**

The activity is proportional to the number present of nuclei of the parent nuclide. This is the same as saying that a *constant fraction* of the number of nuclei of the parent nuclide undergo a transformation per unit time.

In Fig. 16.7, for example, the activity of a quantity of the radioactive

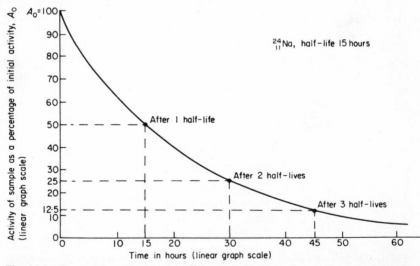

Fig. 16.7   The activity of a sample of $^{24}_{11}Na$ plotted against time, both on linear graph scales. Note the exponential curve with the activity reducing by a factor of ½ every 15 hours.

isotope of sodium, $^{24}_{11}$Na, is plotted against time, both axes of the graph having linear scales. The graph shows that the activity decreases with time and is reduced by a constant factor of one-half every 15 hours (see section 16.5.2 on half-life). This type of relationship, where the activity is reduced by the same factor in successive intervals of time, is known as an **exponential law** and is represented mathematically by the equation:

$$A = A_0 e^{-\lambda t},$$ Eq. 16.1

where A is the activity of the sample after time t,

A_0 is the initial activity of the sample, i.e. when $t = 0$,

e is a mathematical constant, and

λ is the decay constant for the given radionuclide.

It can be shown mathematically that the **decay constant** λ is equal to the *fraction* of the nuclei of the parent radionuclide which undergoes a transformation per unit time.

Note that Eq. 16.1 is analogous to Eq. 14.1 (14.2.3) which represents the transmission of a homogeneous beam of radiation through a medium, and to Eq. 4.7 (4.3.3) which represents the exponential discharge with time of capacitance through resistance.

If the **decay curve** shown in Fig. 16.7 is replotted using a *logarithmic* scale (14.2.3) for the activity axis but retaining a *linear* scale for the time axis, a *straight line* graph is obtained, as shown in Fig. 16.8. This is a useful property of the exponential law.

16.5.2 Half-life. Each radionuclide has its own particular decay rate. In Figs 16.7 (16.5.1) and 16.8 (16.5.2), for example, the activity of a sample of $^{24}_{11}$Na is seen to decrease by one-half every 15 hours. This radionuclide is therefore said to have a **half-life** of 15 hours.

DEFINITION **The half-life of a radionuclide is the time taken for the activity of a sample of that radionuclide to decrease to one-half.**

Half-lives can range from a fraction of a second for some radionuclides to over a million years for others; examples are given in Table 16.1 for a number of radionuclides used in medicine.

The decay constant (Eq. 16.1, 16.5.1) and the half-life for a given radionuclide are related mathematically by the equation:

$$\frac{0 \cdot 693}{\text{half-life}}.$$ Eq. 16.2

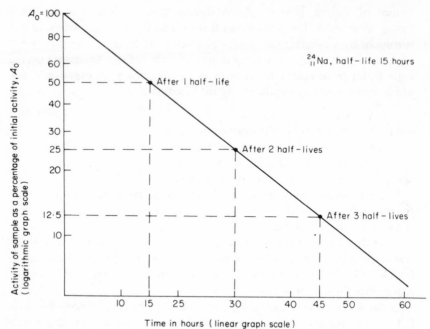

Time in hours (linear graph scale)

Fig. 16.8 (16.5.1). The activity of a sample of $^{24}_{11}$Na plotted on a logarithmic graph scale against time on a linear graph scale. Note that the exponential curve in Fig. 16.7 has become a straight line when plotted on log-linear graph scales.

This is analogous to Eq. 14.2 (14.2.5) which relates the total linear attenuation coefficient of a medium to the half-value layer for a beam of homogeneous radiation.

16.5.3 Average life. It is not possible to say how long the nucleus in *one particular atom* in a sample of a radionuclide will exist before it undergoes a radioactive transformation, but it *is* possible to calculate the **average life** expectancy of a nucleus. This is given by the equation:

$$\text{average life} = \frac{1}{\lambda} = \frac{\text{half-life}}{0\cdot693} \text{,}$$

where λ is the decay constant of the radionuclide (16.5.1).

For $^{24}_{11}$Na, which has a half-life of 15 hours, the average life of a nucleus is

$$\frac{15}{0\cdot693} = 21\cdot6 \text{ hours.}$$

TABLE 16.1 (16.5.2) *Examples of artificial radionuclides used in medicine* (Abridged data)

Radionuclide	Half-life	Particle emitted	Maximum energy of particles, MeV	Gamma-ray photon energy, MeV	Exposure rate constant R m² h⁻¹ Ci⁻¹ (section 16.7)	Uses
$^{3}_{1}$H	12·3 y	β⁻	0·018	no γ	—	⎫ tracer studies
$^{14}_{6}$C	5760 y	β⁻	0·159	no γ	—	⎬
$^{22}_{11}$Na	2·6 y	β+, E.C.*	0·54	(0·51)†, 1·28	1·2	therapy and tracer studies
$^{24}_{11}$Na	15·0 h	β⁻	1·39	1·37, 2·75	1·84	
$^{32}_{15}$P	14·3 d	β⁻	1·71	no γ	—	tracer studies
$^{35}_{16}$S	87·2 d	β⁻	0·167	no γ	—	tracer studies
$^{45}_{20}$Ca	165 d	β⁻	0·254	no γ	—	⎫ tracers for bone studies and other calcium metabolism studies
$^{47}_{20}$Ca	4·7 d	β⁻	0·69, 2·00	0·50, 0·81, 1·31	0·57	⎬
$^{51}_{24}$Cr	27·8 d	E.C.*	—	0·323	0·016	⎫ tracers for blood studies
$^{59}_{26}$Fe	45 d	β⁻	0·27, 0·46	1·10, 1·29	0·64	⎬
$^{58}_{27}$Co	71 d	β+, E.C.*	0·485	(0·51)†, 0·81	0·55	tracer for anaemia studies
$^{60}_{27}$Co	5·26 y	β⁻	0·31	1·17, 1·33	1·32	teletherapy units and therapy sources
$^{131}_{53}$I	8·04 d	β⁻	0·33, 0·61	0·36, 0·64	0·22	therapy and tracer for thyroid studies
$^{132}_{53}$I	2·3 h	β⁻	several from 0·80 to 2·14	many, most abundant 0·52, 0·65, 0·67, 0·78, 0·95	1·18	tracer for thyroid studies
$^{137}_{55}$Cs	30 y	β⁻	0·51, 1·17	0·662 (via ^{137m}Ba)	0·33	teletherapy units
$^{198}_{79}$Au	2·7 d	β⁻	0·96	0·412	0·23	therapy
$^{197}_{80}$Hg	65 h	E.C.*	—	0·077, (0·069)	0·035	tracer for kidney studies

* E.C. = electron-capture process (16.3.4).
† 0·51 = positron-annihilation radiation (14.3.5).

16.5.4.   The curie. The activity of a sample of a radionuclide was defined in section 16.5.1 as the number of nuclear transformations which occur in the sample per unit time. The special unit used for activity is the **curie,** abbreviated to Ci.

DEFINITION   **The curie is that quantity of a radionuclide in which the number of nuclear transformations occurring per second is $3 \cdot 7 \times 10^{10}$.**

The curie was taken originally as the activity of one gram of radium which has approximately $3 \cdot 7 \times 10^{10}$ nuclear transformations per second. It is not an S.I. unit and is being replaced by a new S.I. unit called the **becquerel (Bq)**, which is defined as that quantity of a radionuclide in which one nuclear transformation occurs per second. Note that 1 curie = $3 \cdot 7 \times 10^{10}$ becquerel.

16.6   ARTIFICIAL OR INDUCED RADIOACTIVITY

16.6.1   The production of artificial radionuclides. Sections 16.1.2 and 16.1.3 dealt with *naturally occurring* radioactive substances such as radium and uranium. It is also possible to produce very many *artificial* radionuclides, i.e. radionuclides which do not occur in nature. Most of these are made by bombarding a stable (non-radioactive) nuclide with neutrons inside a **nuclear reactor** at an atomic energy establishment. Inside a reactor there is a very high intensity of neutrons; these interact with the nuclei of stable nuclides placed in the reactor making the nuclei unstable and thus forming radionuclides.

Several different types of nuclear process can occur during the production of radionuclides in a reactor. In one of the simpler processes, a neutron enters a stable nucleus to form an unstable isotope of the same element, e.g.

$$^{23}_{11}\text{Na} + ^{1}_{0}n \rightarrow ^{24}_{11}\text{Na} + \gamma.$$

The stable isotope of sodium, $^{23}_{11}\text{Na}$, becomes a radioactive isotope of the same element, $^{24}_{11}\text{Na}$, with the emission of a gamma photon which represents the excess of energy in the interaction. Subsequently the $^{24}_{11}\text{Na}$ decays with a half-life of 15 hours (Figs. 16.7, 16.5.1 and 16.8, 16.5.2). Note that the unstable isotope has a different ratio of neutrons to protons in the nucleus from that in a nucleus of the stable isotope (2.3).

Other artificial radionuclides are formed by the bombardment of stable nuclides with electrically charged particles such as protons or alpha particles in a **cyclotron** (a large electrical machine for accelerating charged particles to high velocities). Various nuclear reactions can occur including the addition of an alpha particle to a stable nucleus to form an unstable nucleus, e.g.

$$^{52}_{24}Cr + ^{4}_{2}a \rightarrow ^{52}_{26}Fe + 4^{1}_{0}n.$$

The stable isotope of chromium, $^{52}_{24}Cr$, becomes a radioactive isotope of iron, $^{52}_{26}Fe$, with the emission of 4 neutrons. Subsequently the $^{52}_{26}Fe$ decays with a half-life of 8·3 hours.

16.6.2   Specific activity. Although the activity (16.5.1) of a sample is used as a measure of the quantity of the radionuclide present, it does not say anything about the mass or volume of the material in which the radioactive transformations are occurring. The relationship between the activity and the mass of the radioactive material is known as the **specific activity**.

DEFINITION   **Specific activity is the activity per unit mass of the radioactive element or compound.**

It can be expressed in units of curies per kilogram or of curies per mole. A **mole** is the number of grams of the substance numerically equal to its molecular weight (i.e. the average weight of the molecules expressed in atomic mass units, see footnote Table 2.1, 2.1.3).

It is possible to prepare samples of a radionuclide to which no non-radioactive material or **carrier** has been added. Such a sample is said to be **carrier-free**.

16.7   THE EXPOSURE RATE CONSTANT

The exposure rate constant (which replaces the former specific gamma-ray constant) of a radionuclide provides a measure of the exposure rate (15.2.1) at a given distance from a point source of a gamma-ray emitter.

DEFINITION   **The exposure rate constant of a radionuclide is the exposure rate at unit distance from a point source of unit activity.**

It is expressed in units of roentgens per hour at 1 metre from 1 curie $(R\,m^2\,h^{-1}\,Ci^{-1})$ or roentgens per hour at 1 cm from 1 millicurie $(R\,cm^2\,h^{-1}\,mCi^{-1})$. Note that $1\,R\,m^2\,h^{-1}\,Ci^{-1}$ is equivalent to $10\,R\,cm^2\,h^{-1}\,mCi^{-1}$.

The gamma rays from a point source obey the inverse square law in a non-absorbing medium in the same way as does all electromagnetic radiation

(9.1.4). Consequently, if the exposure rate constant for a radionuclide is known, the inverse square law can be used to calculate the exposure rate at any given distance from a geometrically small source of any activity of that radionuclide.

16.8 THE USES OF RADIONUCLIDES IN MEDICINE

In addition to naturally occurring radium (16.1.3), many of the artificial radionuclides have found applications in medicine where their varied chemical properties as well as their differing radiation qualities have been of use. Table 16.1 (16.5.2) lists a few of these radionuclides with their half-lives, emissions, and uses in medicine.

17 Radiological protection

17.1 HISTORICAL INTRODUCTION

Man has always been exposed to background radiation. This has two components: first, radiation from the radioactive substances in his surroundings and within his body, and second, cosmic radiation which reaches the earth from space. It was not until after the discovery of X rays in 1895 and of radioactivity in 1896, however, that the need for radiological protection was recognized. Early workers with X rays and radioactive substances soon realized that the radiations can cause burns and other injuries, and that doses of radiation too small to cause immediate injury can result in harmful effects, such as tumours and leukaemias, many years after exposure. These effects which are produced in an individual's body are called **somatic** effects (14.1.2). In addition, radiation can cause **genetic** effects; these result from mutations (changes) of the genes in the chromosomes of reproductive cells. The chromosomes govern the inherited characteristics of individuals in subsequent generations.

Early efforts to promote radiological protection were made by the Röntgen Society in 1916. They recommended the use of protective shields, inspection of equipment, restricted working hours, and special medical examinations including blood tests. Subsequently the British X-Ray and Radium Protection Committee presented its first radiological protection measures in 1921.

Current practice in radiological protection is based on the various recommendations published by the International Commission on Radiological Protection (I.C.R.P.), with each country producing national Codes of Practice

247

and legislation which interpret and elaborate these recommendations. In the United Kingdom, hospitals are subject to the *Code of Practice for the Protection of Persons against Ionizing Radiations arising from Medical and Dental Use*, issued by the Department of Health and Social Security. The Code applies to the use of ionizing radiations arising from all forms of medical and dental practice and from allied research in hospitals; it sets out basic principles for radiation control and general guidance for good practice. The sections of the Code which relate to radiographers are summarized below (17.3).

In 1974, the Health and Safety at Work etc. Act was passed in the United Kingdom. This Act is having far-reaching effects on all aspects of safety and will probably cause some changes in the arrangements for radiological protection in hospitals.

17.2 MAXIMUM PERMISSIBLE DOSES

For practical purposes, it is desirable to determine the maximum doses of radiation which can be assumed to constitute a negligible hazard to people. The International Commission on Radiological Protection has therefore defined a **permissible dose** for an *individual.*

DEFINITION A **permissible dose is that dose, accumulated over a long period of time or resulting from a single exposure, which, in the light of present knowledge, carries a negligible probability of severe somatic or genetic injuries; furthermore, it is such a dose that any effects which ensue more frequently are limited to those of a minor nature that would not be considered unacceptable by the exposed individual or by competent medical authorities.**

Also, when considering the general public, i.e. a large number of people as opposed to the relatively small number of individuals who are radiation workers, the doses must be further limited so that they do not result in an unacceptable burden of genetic changes in future generations.

On this basis, the Code of Practice gives values for the **maximum permissible doses** for radiation workers and for other persons. These are summarized in Table 17.1. The maximum permissible doses for the whole body, the red bone-marrow and the gonads are less than those for single organs or for the extremities such as the hands. For example, exposure of the red bone-marrow (the blood-forming organ) is considered to be more hazardous because of the possibility of leukaemia.

TABLE 17.1 (17.2) *Maximum permissible doses for designated radiation workers and for members of the public**

Exposed part of body	Designated radiation workers (17.3.1) over 18 years of age	Members of the public
Whole body, red bone-marrow, and gonads	Cumulative dose = $5(N-18)$ rems† where N is the age in years;‡ 3 rems/13 weeks	0.5 rems/year
Bone, thyroid, and skin of whole body	15 rems/13 weeks; 30 rems/year	3 rems/year
Hands, forearms, feet, and ankles	40 rems/13 weeks; 75 rems/year	7·5 rems/year
Any single organ (excluding red bone-marrow gonads, bone, thyroid, and skin of whole body)	8 rems/13 weeks; 15 rems/year	1·5 rems/year

* Based on the figures given in the *Code of Practice for the Protection of Persons against Ionizing Radiations arising from Medical and Dental Use*, 1972.

† The **rem** is the unit of **dose equivalent** used for protection purposes. The **dose equivalent** is the product of the absorbed dose in rads and a quality factor which takes into account the effectiveness of the radiation to cause biological damage. For X rays, the quality factor is 1; an absorbed dose of 1 rad of X rays is therefore equivalent to 1 rem.

‡ The formula gives the *cumulative* maximum permissible dose to the whole body, red bone-marrow, and gonads that a radiation worker may receive; it is equivalent (arithmetically) to an average dose of 5 rems for every year that the individual's age exceeds 18 years or, in terms of weekly dose, an average of 100 millirems per week.

Provided that the *cumulative* dose permitted by the formula is not exceeded, a radiation worker may receive up to 3 rems during a 13-week period except in the case of abdominal exposure of women of reproductive age for which a limit of 1·3 rems in a 13-week period is set. After diagnosis of a pregnancy, the average dose to the fetus should not exceed 1 rem during the remaining period of the pregnancy.

With the introduction of the gray (15.4.1) as the S.I. unit of absorbed dose, the rem is being replaced by a corresponding S.I. unit of dose equivalent called the **sievert (Sv)**. Note that 1 sievert = 100 rem.

The above maximum permissible doses do not apply to patients because the benefits of a particular radiation procedure will probably far outweigh the small risk involved. It is nevertheless desirable to limit the exposure of patients to the minimum value consistent with the medical requirements; various techniques are recommended in the Code of Practice to ensure this (17.3.2, 17.3.3, 17.3.4).

17.3 THE CODE OF PRACTICE

17.3.1   General. Radiographers must be familiar with the relevant sections and appendices in the *Code of Practice for the Protection of Persons against Ionizing Radiations arising from Medical and Dental Use.* They must also know the **Local Rules** required by the Code of Practice, which set out the detailed procedures for protection in force in their particular department of the hospital, and, if involved in dental radiology, they should be familiar with the booklet, *Radiological Protection in Dental Practice*, issued by the Department of Health and Social Security in 1975.

The ultimate responsibility for protection lies with the Controlling Authority for the hospital. The Code of Practice goes on to recommend an administrative organization in which responsibility is divided among the Radiological Protection Adviser (who must be a physicist), the Supervisory Medical Officer, Departmental Radiological Safety Officers, Heads of Department or Clinicians, and the individual members of staff.

The Controlling Authority must determine which members of staff are to be **designated** as radiation workers. **Designated persons** are those whose work is associated with radioactive substances or apparatus emitting ionizing radiations and who are liable to receive doses in excess of those given for 'other staff' in Table 17.1. These workers must have medical examinations, including blood examinations, and must wear personal radiation monitors such as film-badge dosemeters. They may then be permitted, if necessary, to receive radiation doses up to the higher levels given in Table 17.1.

17.3.2   Diagnostic uses of X rays. Protection in diagnostic radiology is designed to achieve two ends: first, to limit the exposure of the patient to the minimum that will produce a satisfactory result, and second, to protect all other persons from the useful beam and from scattered and leakage radiation. The first is achieved by:

Limitation of field size to the minimum area necessary by the use of diaphragms and cones, especially when examining children and pregnant women.

Filtration of the beam by at least 1 or 2 mm of aluminium (14.5).

Directing the beam, whenever possible, so that the gonads are not irradiated by the useful beam.

Use of gonad shields, if practicable, whenever there is risk of an appreciable dose to the gonads, e.g. in examinations of the abdomen of the female patient and of the chest of the male or female.

250

Careful preparation of the patient so that the need for a second examination is avoided.

Use of the highest speed film emulsions and intensifying screens consistent with good radiographic detail.

Use of automatic timing devices, such as photoelectric or ionization chamber timers (13.4.2), to minimize the need for second examinations.

Dark adaptation for at least 10 minutes before the traditional type of fluoroscopic examination.

Use of automatic time-limiting switches or clocks (13.4.1) with fluoroscopy machines.

Use of image intensification for fluoroscopy whenever possible.

The second is achieved by:

Presence only of essential staff in the X-ray room during radiological examinations.

Use of protective aprons or protective screens to wear or to stand behind.

Provision of shielding in the tube housing so that the 'leakage radiation' through the housing does not exceed 10 milliroentgens in one hour* at 1 metre distance from the focus.

Use of immobilizing devices for children and adults requiring support during an examination.

Provision of shielding in the walls of X-ray rooms to reduce the doses received by persons in adjacent rooms.

Only essential radiological examinations should be conducted on pregnant women and particular care should be taken to avoid irradiation of the fetus. Chest examinations of pregnant women should not be carried out by mass miniature techniques, but with full size films (or with image intensifiers and 70- or 100-mm cameras) and with strict limitation of field size.

Because of the possibility of an unrecognized pregnancy and the extreme sensitivity to radiation of the embryo in the early stages, the International Commission on Radiological Protection recommends the following: that all radiological examinations of the lower abdomen and pelvis of women of reproductive capacity, that are not of importance in connexion with the immediate illness of the patient, should be performed only during the **ten-day**

---

*This value may be changed back to the earlier recommendation of 100 milliroentgens in one hour because of the technical difficulties in achieving the lower figure.

interval following the onset of menstruation, when pregnancy is improbable. In the case of a child, there should be a positive indication of the need for a radiological examination before such an examination is performed, in order to avoid unnecessary exposure.

The exposure of staff in a radiodiagnostic department is monitored usually by film-badges (17.5.2) and records are kept of the doses received by each individual. The film-badges are normally worn under any protective aprons used; therefore they measure essentially the exposure of the whole body. The doses received have to be compared with the maximum permissible cumulative dose of $5(N - 18)$ rems for a person N years of age (Table 17.1), which is equivalent arithmetically to an average weekly dose of 100 millirems. If the maximum permissible dose for the period of issue of a monitoring film (usually 2 or 4 weeks) is exceeded, the Departmental Radiological Safety Officer would normally investigate the reason for the excessive exposure in order to prevent its recurrence. Among the causes of excessive exposure of members of staff are inadvertent irradiation by the useful beam, too great a time spent near a fluoroscopy unit, and failure to use a protective screen or apron when appropriate.

17.3.3 Therapeutic uses of X rays. Because of the high exposure rates involved in radiotherapy, there must be a rigid procedure for operating the equipment which ensures that neither the patient nor the staff is accidentally overexposed. Except for machines used for superficial therapy and operating below 100 kVp, all persons other than the patient must be outside the treatment room during the exposure and the entrance to the room must be fitted with an interlock which automatically shuts off the machine if the room is entered. The room must be so constructed that the walls provide adequate shielding and any observation windows (some of which are made of lead-glass) must provide at least the same degree of protection as the walls in which they are set. To reduce the exposure rate at the entrance, access is often through a maze so that the only radiation reaching the entrance has been greatly attenuated by repeated scattering and little or no shielding is required in the door. Sometimes mechanical limit stops are fitted on the tube mounting to prevent the useful beam from being sent in certain directions; consequently the shielding provided by the walls in these directions is required to protect personnel against leakage and scattered radiation only.

The Code of Practice states that an automatic timer or dosemeter must be provided to terminate the treatment after a pre-set time or dose. Also, all

X-ray machines used for therapy should be calibrated at intervals of not more than 4 weeks by a physicist.

The leakage radiation from the tube housing should not exceed 1 roentgen/hour at 1 metre from the focus or 30 roentgens/hour at any point accessible to the patient at a distance of 50 mm from the surface of the housing.

17.3.4   Use of radioactive substances. Radioactive substances are used in diagnosis and in therapy. In diagnosis, their main use is in tracer investigations to follow the metabolism of a given chemical substance administered to a patient or to check the functioning of an organ. Two examples of this are the use of calcium-47 to study the growth of bone and iodine-132 to measure thyroid function. In therapy, they are used: (1) as large sealed sources for teletherapy, for example in cobalt-60 and caesium-137 units, (2) as (physically) small sealed sources such as radium needles and gold-198 grains for implantation in tumours, and (3) as unsealed sources such as solutions of iodine-131 for thyroid ablation and suspensions of colloidal gold-198 for treatment of malignant pleural or peritoneal effusions.

If the substances emit gamma radiation, the protection problems are similar to those associated with X rays and the shielding required is discussed in section 17.4. Radioactive substances also emit either alpha particles or beta particles (16.3.1, 16.3.2). Alpha particles are very easily absorbed and all are completely stopped by about 120 mm of air or *about one-tenth* of a millimetre of materials such as water, Perspex, or soft tissue. Almost any container therefore provides adequate shielding against alpha particles.

Beta particles have a range of up to several metres in air or a few tens of millimetres in Perspex or soft tissue. A thickness of 10 or 20 mm of Perspex or the glass wall of a bottle therefore usually provides good if not complete shielding. Note, however, that if the beta particles (fast-moving electrons) are of high energy, the Bremsstrahlung (X radiation, 10.2.1, 10.3(iv)) produced when they are stopped in the shielding may have a high intensity. Additional shielding to attenuate the Bremsstrahlung may be required, especially if the source is of high activity.

The precautions to be taken with a teletherapy unit are similar to those for a therapeutic X-ray machine. The main difference between the two types of installation is that an X-ray beam can be switched off but a heavy shutter or equivalent arrangement has to be used to interrupt the gamma-ray beam. The Code of Practice recommends limits under various conditions for the leakage radiation through the housing containing the source in a teletherapy unit. For example, with the unit 'off', the maximum exposure rate at 1 metre from the source, in any direction, should not exceed 10 milliroentgens/hour. An additional hazard with a teletherapy unit is the possibility of the radioactive substance itself leaking out of the capsule in which it was sealed; therefore periodic tests must be made for contamination of the surfaces of the unit.

When radioactive substances are dispensed and administered for tracer studies or for therapy, precautions must be taken by members of staff against accidental ingestion, inhalation or absorption 'through the skin. Once inside the body, there is no shielding against alpha or beta particles and, as their ranges are small in tissue, all the energy of a particle is delivered to a small mass of tissue with a consequent increase in the likelihood of damage. This is particularly important in the case of alpha particles as they have very short ranges in tissue. Also, certain organs in the body take up various radioactive substances because of the chemical properties of the substances; the radioactivity is therefore concentrated into a relatively small mass and the hazard consequently increased, e.g. the uptake of iodine-131 by the thyroid. When this happens, the organ at greatest risk is known as the **critical organ**. These considerations lead to the specification of the **maximum permissible body-burdens** (for the various radioactive substances) which result in the maximum permissible dose being received by the relevant critical organ.

17.4 PROTECTIVE MATERIALS FOR X AND GAMMA RADIATION

17.4.1 The choice of materials. The dose received by a person can be limited by restricting the duration of the exposure and by increasing the distance between the person and the source (inverse square law, 9.1.4) but often it is necessary to reduce the exposure rate further by means of a protective barrier. The material chosen for the barrier should have a high linear attenuation coefficient (14.2.4) over the range of energies concerned. The *linear* attenuation coefficient varies as the cube of the atomic number multiplied by the density if the attenuation is by the photoelectric process but is independent of the atomic number and varies only as the density if the attenuation is by the Compton process (14.3.3, 14.3.4, 14.3.7). Lead, with its high density of $1 \cdot 1 \times 10^4$ kg m^{-3} (11 g cm^{-3}) and high atomic number of 82, is often used up to about 300 kVp because the photoelectric process is predominant and the high atomic number results in a large attenuation. For X rays generated at higher voltages and for high energy gamma rays such as those from cobalt-60 (1.17 and 1.33 MeV), the Compton process predominates and a high atomic number is no longer advantageous. Also, the thickness of lead required at high energies would prove too costly for extensive shielding. It is cheaper and easier to use adequate thicknesses of materials such as concrete (density $2 \cdot 3 \times 10^3$ kg m^{-3} ($2 \cdot 3$ g cm^{-3})) incorporated into the building structure. For a cobalt-60 treatment room, for example, about $1 \cdot 2$ metres of concrete is necessary to shield against the useful beam but less for the walls receiving leakage and scattered radiation only.

The attenuation in walls can be increased by modifying ordinary building materials. The density of concrete can be increased by the addition of barium sulphate or iron pellets. For the walls of diagnostic X-ray rooms, where the photoelectric process makes an important contribution to the attenuation, the calcium in ordinary plaster can be replaced by barium which has a higher atomic number.

For special applications, particularly where compact shielding is required as in the head of a teletherapy unit or in the diaphragms for limiting field size, high density materials such as tungsten and uranium, both with densities of about $1 \cdot 9 \times 10^4$ kg m^{-3} (19 g cm^{-3}), are used even though the uranium is slightly radioactive. Also, because uranium has a very high atomic number, the photoelectric process is significant up to higher energies than for lead and it is sometimes used effectively in teletherapy units containing caesium-137 (gamma-ray energy $0 \cdot 662$ MeV).

17.4.2 Lead equivalent. The effectiveness of a protective barrier is often described in terms of its **lead equivalent**.

DEFINITION **The lead equivalent of a barrier is that thickness of lead affording the same degree of protection under specified conditions of irradiation as the barrier.**

The lead equivalent of a barrier made of a material, such as the lead-glass in a fluoroscopy screen or the lead-rubber in a protective apron, which attenuates the radiation essentially by its *lead content,* varies very little with the energy of the radiation. In contrast, the lead equivalent of a protective barrier *which does not contain lead,* such as barium plaster or concrete, does vary with the quality of the radiation because the attenuation coefficients for the material and for lead vary by different amounts as the quality of the radiation changes.

17.5   RADIATION MONITORING

17.5.1   Monitoring. The continuing effectiveness of radiation protection measures must be checked by monitoring the doses received by personnel and the exposure rates at various sites, especially after any modifications have been made to the sources of radiation or to the shielding. Besides the Geiger counters and scintillation detectors described in sections 15.5.4 and 15.5.5, several other types of radiation detector are used for monitoring, including the following.

17.5.2   Film-badges (15.5.2). The film-badge consists of a photographic film with two coatings of emulsion enclosed in a thin light-tight wrapper and mounted in a special holder (Fig. 17.1). Radiation produces a latent image on

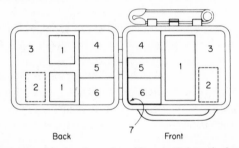

Back   7   Front

Fig. 17.1 The film-badge showing the various filters built into the back and the front of the holder. 1. window; 2. thin plastics (50 mg cm^2); 3. thick plastics (300 mg cm^2); 4. 0.040 inch thick Dural; 5. 0.028 inch cadmium + 0.012 inch lead; 6. 0.028 inch tin + 0.012 inch lead; 7. 0.012 inch lead edge shielding.

the film and, after development, the photographic density of the resultant blackening is measured using a densitometer. In order to relate the density of blackening to the dose received, the type of radiation (whether beta or X radiation) and the energy of the radiation must be known (i.e. the sensitivity of the film depends on the type and energy of the radiation). Often the conditions of irradiation of the film are not precisely known and so filters are built into the holder so that the type and energy of the radiation can be deduced from the relative values of the photographic densities produced under the various filters. The filters commonly include areas of thin plastics, thicker plastics, Duralumin (an aluminium–copper alloy commonly called Dural), combined layers of cadmium plus lead and tin plus lead, and an open window with no filter. Low-energy X radiation and beta particles, for example, will pass through the open window but will be attenuated by the Dural filter. Higher energy X radiation will show little difference in blackening under these two areas but will be attenuated by the tin–lead filter.

The two emulsions on the film, one fast and the other slow, enable the film-badge to be used to measure the small doses encountered under good working conditions and also to record an accidental high exposure. The fast emulsion has a range for X-ray exposures from about 5 milliroentgens to 1 roentgen. A high exposure would completely blacken the fast emulsion but this can be stripped off leaving the slow emulsion with a range from 1 to 100 roentgens.

The film badge dosemeter therefore provides a simple, robust, permanent record for a wide range of exposures with an indication of the type and energy of the radiation involved. A film is usually worn for 2 to 4 weeks for personnel monitoring. The main disadvantages are that no immediate indication of the exposure is given and the latent image is affected if the film is heated, for example, by placing it near a heating radiator.

17.5.3 Ionization chambers with air-equivalent walls are described in sections 15.2.3 and 15.2.4. Small chambers of this type can be used to measure the exposure to X or gamma radiation over a wide range of energies because of their air-equivalence but, in contrast to a film dosemeter, no indication is given of the type or energy of the radiation involved and no permanent record of the exposure is produced. For protection purposes, the small Medical Research Council condenser chamber type BD11 (15.2.4) is often used for exposure up to about 0·5 roentgen. It is charged from a separate unit which also contains an electronic form of electrometer for measuring the charge.

In some situations it is necessary to have an immediate indication of the dose accumulated; this is provided in the direct-reading quartz fibre electrometer dosemeter (commonly called a 'fountain-pen' dosemeter). This consists of an ionization chamber connected to a quartz fibre electrometer which is viewed through a lens system built into the instrument. The chamber is charged initially using a separate unit but, during the exposure, the loss of charge, which is proportional to the accumulated dose, can be read directly by viewing the deflexion of the quartz fibre through the lens system. The dosemeter is about the size and shape of a fountain pen and is available in several models with ranges having full-scale readings of 0·2 to 50 roentgens.

17.5.4   Thermoluminescent dosemeters (15.5.3). Certain crystalline powders such as lithium fluoride possess the property of **thermoluminescence**, that is they store energy during exposure to ionizing radiation and subsequently release it as light when heated.

To measure dose, the powder is usually placed in a small plastic container. After exposure to the radiation, the powder is removed from the container and heated in an instrument known as a 'reader'. This measures the amount of light released which is proportional to the radiation dose received by the powder.

Thermoluminescent dosemeters are capable of measuring a wide range of exposures from a few milliroentgens to hundreds of roentgens; they are therefore used for protection work and for measuring doses received in therapy. Unlike the film-badge which provides a permanent record of exposure, the information stored in a thermoluminescent powder can be read out only once and no permanent record remains. However, thermoluminescent dosemeters can be processed automatically which is a considerable advantage when large numbers are involved. In consequence, they may replace film-badges for routine monitoring in the future.

Index